Innovative Manufacturing Engineering and Energy IMANEE2025

International Conference on Innovative Manufacturing Engineering and Energy IMANEE2025, May 15th-16th 2025, Iasi, Romania

Editor

Marius-Andrei Mihalache[1]

[1]Department of Machine Manufacturing Technology (https://www.tcm.cmmi.tuiasi.ro/member/andrei-mihalache/), Faculty of Machine Manufacturing and Industrial Management (https://cmmi.tuiasi.ro/) from Gheorghe Asachi Technical University of Iasi, Romania (https://tuiasi.ro

Peer review statement

All papers published in this volume of "Materials Research Proceedings" have been peer reviewed. The process of peer review was initiated and overseen by the above proceedings editors. All reviews were conducted by expert referees in accordance to Materials Research Forum LLC high standards.

Published under License by **Materials Research Forum LLC**
Millersville, PA 17551, USA

Published as part of the proceedings series
Materials Research Proceedings
Volume 61 (2026)

ISSN 2474-3941 (Print)
ISSN 2474-395X (Online)

ISBN 978-1-64490-398-8 (Print)
ISBN 978-1-64490-399-5 (eBook)

This book contains information obtained from authentic and highly regarded sources. Reasonable efforts have been made to publish reliable data and information, but the author and publisher cannot assume responsibility for the validity of all materials or the consequences of their use. The authors and publishers have attempted to trace the copyright holders of all material reproduced in this publication and apologize to copyright holders if permission to publish in this form has not been obtained. If any copyright material has not been acknowledged please write and let us know so we may rectify in any future reprint.

Distributed worldwide by

Materials Research Forum LLC
105 Springdale Lane
Millersville, PA 17551
USA
https://mrforum.com

Manufactured in the United State of America
10 9 8 7 6 5 4 3 2 1

Table of Contents

GHEORGHE ASACHI TECHNICAL UNIVERSITY OF IASI, ROMANIA

The INGENIUM European University Consortium represents a strategic alliance of leading technical and comprehensive universities across Europe, established to advance excellence in education, research, innovation, and societal engagement. Operating within the framework of the European Universities Initiative, INGENIUM brings together institutions that share a common commitment to academic quality, interdisciplinarity, openness, and long-term institutional transformation. The consortium functions as an integrated academic ecosystem rather than a simple partnership, fostering structural cooperation and co-creation across national, cultural, and disciplinary boundaries.

Within this context, the Gheorghe Asachi Technical University of Iași (TUIASI), Romania, holds a well-defined and increasingly visible position as a full member of the INGENIUM consortium. TUIASI contributes with its strong tradition in engineering, applied sciences, and technological innovation, complementing the academic profiles of the other partner universities. Its membership reinforces the technical and applied dimension of INGENIUM, ensuring that engineering education, digital transformation, sustainability, and industry-oriented research remain central pillars of the alliance.

The overarching aims and goals of INGENIUM are aligned with the creation of a truly European university: integrated education pathways, joint research agendas addressing major societal challenges, shared innovation ecosystems, and a strong emphasis on inclusiveness and academic mobility. The consortium focuses on key strategic directions such as sustainable development, digitalization, advanced manufacturing, smart technologies, and societal resilience. Through these directions, INGENIUM seeks not only to harmonize curricula and research infrastructures but also to shape a common academic culture rooted in excellence, integrity, and impact.

TUIASI operates under the patronage of INGENIUM through a wide range of already implemented and ongoing activities. These include participation in joint educational initiatives such as common modules, blended intensive programs, and co-developed learning outcomes, as well as involvement in thematic research working groups and interdisciplinary task forces. Academic and administrative staff from TUIASI have engaged in consortium-level governance structures, mobility schemes, capacity-building workshops, and best-practice exchanges, contributing actively to the institutional development objectives of INGENIUM. Further activities are planned and underway, encompassing expanded joint degree concepts, shared digital platforms, collaborative doctoral frameworks, and reinforced links with industry and regional stakeholders.

A central dimension of the INGENIUM–TUIASI collaboration is the broad spectrum of opportunities it offers to students. Students within the consortium benefit from enhanced academic mobility, access to diverse educational resources, exposure to multicultural learning environments, and participation in joint courses and projects that transcend traditional institutional boundaries. These opportunities foster transversal skills, research-based learning, and employability, while also cultivating a strong sense of European academic identity. Beyond enrolled students, early-career researchers, doctoral candidates, and teaching staff equally benefit from shared infrastructures, mentoring networks, and collaborative research opportunities embedded within INGENIUM.

At an institutional level, the INGENIUM affiliation significantly enhances the international visibility and strategic positioning of TUIASI. Participation in a prestigious European University alliance elevates TUIASI's profile as a competitive, forward-looking technical university and strengthens its role as a reference point within the Romanian higher education landscape. Through its active engagement and concrete contributions, TUIASI emerges as a pillar of good practice, demonstrating how Romanian universities can successfully integrate into high-impact European academic structures while preserving their institutional identity and strengths.

The INGENIUM consortium provides a robust framework through which TUIASI advances its mission in education, research, and innovation. The collaboration generates tangible benefits for students, academics, and society at large, while positioning TUIASI as a key contributor to the future of European higher education and as a model to be followed by other universities in Romania and beyond.

Preface

The Innovative Manufacturing Engineering and Energy (IManEE) international conference is conceived as a high-level scientific forum dedicated to advancing knowledge, fostering collaboration, and shaping future directions in manufacturing and energy-related engineering disciplines. Its overarching aim is to create a rigorous yet inclusive environment in which researchers, academics, industry practitioners, and early-career scientists can exchange validated scientific results, discuss emerging challenges, and present innovative solutions that respond to the evolving demands of modern manufacturing systems.

The 29th edition of the Innovative Manufacturing Engineering & Energy International Conference (IManEE 2025) was held in Iasi, Romania being organized by the Machine Manufacturing Technology Department from the Faculty of Machine Manufacturing and Industrial Management. The faculty is one of the eleven that comprise the Technical University Gheorghe Asachi from Iasi. The above-mentioned department is the main organizer of each yearly edition, alongside other peers that include but are not limited to the School of Mechanical Engineering of the National Technical University of Athens, Greece, the Technical University of Moldova from Chisinau, Republic of Moldova, as well as similar departments from Romanian universities such as Brasov, Bucuresti, Cluj, Galati, and Pitesti.

A central objective of IManEE is to facilitate the dissemination of state-of-the-art research in key domains that define contemporary and future industrial paradigms. By focusing on topics such as *Additive Manufacturing, Advanced Manufacturing Processes, Industry 4.0 / 5.0, Smart Manufacturing, Mechanical Engineering, Mechatronics and Robotics*, the conference seeks to capture both technological depth and systemic integration, emphasizing the convergence between digitalization, automation, human-centric design, and sustainable engineering practices.

Another major aim is to encourage interdisciplinary dialogue across traditionally distinct fields. IManEE promotes ideas between materials science, process engineering, digital tools, control systems, artificial intelligence, and robotics, recognizing that breakthrough innovations increasingly emerge at the intersection of disciplines. The conference provides a structured platform for exploring how cyber–physical systems, data-driven decision-making, and intelligent automation can be effectively embedded into manufacturing environments while maintaining robustness, flexibility, and energy efficiency. The conference has since become a significant platform for regular participants to exchange ideas and foster collaborations.

This year's conference attracted researchers from countries in the EU, including Bulgaria, Germany, Greece, Hungary, Spain, the Czech Republic, Poland, Romania, the Republic of Moldova, and Ukraine as authors or reviewers.

Four plenary speeches on compelling topics in manufacturing engineering and energy enriched the conference and provided a foundation for meaningful discussions. The plenary sessions featured distinguished members of the academic community. These included Prof. PhD. Eng. Alexandru SOVER from the Ansbach University of Applied Science, Germany, who delivered a presentation on *"How small can a functional screw be? Challenges of additive manufacturing of micro parts"*. Professor Gabriel FRUMUȘANU from the "Dunărea de Jos" University of Galați, Romania, explored *"Estimation of the manufacturing activity performance by digital modeling and comparative assessment"*. On the following day, Professor Fabrizio QUADRINI from Università degli Studi di Roma Tor Vergata, Italy, discussed *"Innovative Materials and Sustainable Technologies at the Space Frontier"*. The event featured also a young researcher keynote speech. It was delivered by Dr. Vasile ERMOLAI from the Gheorghe Asachi Technical University of Iasi, Romania, who presented the *"Multi-material 3D printed polymers. Influencing factors, design considerations and implementation"*.

The success of the conference was made possible through the support of all those involved. Special thanks are extended to the conference organizers, members of the Scientific Committee, reviewers, and, most importantly, all the participants, whose collective efforts contributed to the success of IManEE 2025. A further objective is to support the development of the next generation of researchers and engineers. Through open scientific discussion, constructive peer feedback, and exposure to international research trends, IManEE provides an enabling framework for doctoral candidates and early-career researchers to refine their work, expand professional networks, and engage with established experts in their field of interest.

Ultimately, the IManEE international conference aims to build a sustainable and evolving scientific community, united by a shared commitment to innovation, quality, and societal impact. By continuously addressing emerging manufacturing paradigms and energy challenges, IManEE positions itself not only as a venue for presenting results but as a catalyst for long-term collaboration and future-oriented research, encouraging participants to return to subsequent editions with new ideas, advanced studies, and strengthened international partnerships.

Within the *Innovative Manufacturing Engineering and Energy international conference*, endeavors are part of a much greater effort started by predecessors that has formed and built the IMANEE community we see today. Without their vision and hardwork none of it would be possible. Therefore, thanks to all collaborators from Romania and abroad are mandatory. A special thank you goes to Professor Emeritus Laurențiu Slătineanu, who honors the event with his academic standing and his impressive professional background, giving most-valued advices as he pushes towards excellence in every sense.

As part of such a big team comprised of colleagues, friends, and well-established professionals from various domains, we all agree that our pursuit of academic excellence is a declaration of our commitment to further achieve recognition and validation from the scientific community we are all part of.

Thank you.

Associate Professor Marius Andrei Mihalache,
Gheorghe Asachi Technical University of Iasi
Faculty of Machine Manufacturing and Industrial Management
Machine Manufacturing Technology Department
Iasi, Romania

Committees

Honorary Committee
Vasile MERTICARU, Gheorghe Asachi Technical University of Iasi, Romania
Florin NEGOESCU, Gheorghe Asachi Technical University of Iasi, Romania
Laurenţiu SLĂTINEANU, Gheorghe Asachi Technical University of Iasi, Romania
Angelos MARKOPOULOS, National Technical University of Athens, Greece
Panagiotis KYRATSIS, University of Western Macedonia, Greece
Sergiu MAZURU, Technical University of Moldova, Moldova Republic
Eduard Laurentiu NITU, University of Pitesti, Romania

Scientific Committee
Adriana MUNTEANU, Gheorghe Asachi Technical University of Iasi, Romania
Agathoklis KRIMPENIS, Hellenic Mediterranean University, Greece
Ágota BÁNYAI, University of Miskolc, Hungary
Alexandru Cătălin FILIP, Transilvania University of Brasov, Romania
Alexandru KECHAGIAS, University of Thessaly, Greece
Alexandru SOVER, Ansbach University of Applied Sciences, Germany
Alexei TOCA, Technical University of Moldova, Moldova Republic
Anamaria FEIER, Politehnica University Timisoara, Romania
Anastasios TZOTZIS, University of Western Macedonia, Greece
Andrei Marius MIHALACHE, Gheorghe Asachi Technical University of Iasi, Romania
Angelos MARKOPOULOS, National Technical University of Athens, Greece
António GONÇALVES-COELHO, NOVA School of Science and Technology, Portugal
António José Freire MOURÃO, Nova University Lisbon, Portugal
António MOURÃO , Universidade NOVA de Lisboa, Portugal
Apostolos KORLOS, International Hellenic University, Greece
Arik SADEH, Holon Institute of Technology, Israel
Athanasios MANAVIS, University of Western Macedonia, Greece
Aurel Mihail TITU, Lucian Blaga University of Sibiu, Romania
Aurel TULCAN, Politehnica University Timisoara, Romania
Bogdan Alexandru CHIRIŢĂ, Vasile Alecsandri University of Bacau, Romania
Bruno RADULESCU, Gheorghe Asachi Technical University of Iasi, Romania
Calin NEAMTU, Technical University of Cluj-Napoca, Romania
Carmen BUJOREANU, Gheorghe Asachi Technical University of Iasi, Romania
Carmen Ema PANAITE, Technical University Gheorghe Asachi of Iasi, Romania
Carol SCHNAKOVSZKY, Vasile Alecsandri University of Bacău, Romania
Catalin DUMITRAS, Gheorghe Asachi Technical University of Iasi, Romania
Catalin FETECAU, Dunărea de Jos University of Galati, Romania
Catalina MAIER, Dunărea de Jos University of Galati, Romania
Christophe SAUVEY, Université de Lorraine, France
Claudia-Emilia GIRJOB, Lucian Blaga University of Sibiu, Romania
Constantin BUNGAU, University of Oradea, Romania
Constantin CARAUSU, Gheorghe Asachi Technical University of Iasi, Romania
Costel MIRONEASA, Stefan cel Mare University of Suceava, Romania
Cristian TURC, Politehnica University Timisoara, Romania
Csaba FELHŐ, University of Miskolc, Hungary

Tetiana ROIK, National Technical University of Ukraine, KPI Kyiv, Ukraine
Vasile ERMOLAI, Gheorghe Asachi Technical University of Iasi, Romania
Vasile MERTICARU, Gheorghe Asachi Technical University of Iasi, Romania
Vasilios SPITAS, National Technical University of Athens, Greece
Vasilis DIMITRIOU, Hellenic Mediterranean University, Greece
Vassilios KAPPATOS, Center for Research and Technology, Greece
Victorița RĂDULESCU, University Politehnica of Bucharest, Romania
Viorel BOSTAN, Technical University of Moldova, Republic of Moldova
Viorel PĂUNOIU, Dunarea de Jos University of Galati, Romania
Virgil-Gabriel TEODOR, Dunarea de Jos University of Galati, Romania

Organizing Committee - TUIASI
Constantin CARAUSU, Gheorghe Asachi Technical University of Iasi, Romania
Dumitru NEDELCU, Gheorghe Asachi Technical University of Iasi, Romania
Eugen AXINTE, Gheorghe Asachi Technical University of Iasi, Romania
Florin NEGOESCU, Gheorghe Asachi Technical University of Iasi, Romania
Gheorghe NAGIT, Gheorghe Asachi Technical University of Iasi, Romania
Laurentiu SLATINEANU, Gheorghe Asachi Technical University of Iasi, Romania
Lucian TABACARU, Gheorghe Asachi Technical University of Iasi, Romania
Margareta COTEATA, Gheorghe Asachi Technical University of Iasi, Romania
Marius Ionut RIPANU, Gheorghe Asachi Technical University of Iasi, Romania
Mihai BOCA, Gheorghe Asachi Technical University of Iasi, Romania
Oana DODUN-DES-PERRIERES, Gheorghe Asachi Technical University of Iasi, Romania
Petru DUSA, Gheorghe Asachi Technical University of Iasi, Romania
Simona MAZURCHEVICI, Gheorghe Asachi Technical University of Iasi, Romania
Teodor Daniel MÎNDRU, Gheorghe Asachi Technical University of Iasi, Romania
Vasile ERMOLAI, Gheorghe Asachi Technical University of Iasi, Romania
Vasile MERTICARU, Gheorghe Asachi Technical University of Iasi, Romania

Organizing Committee - UGAL
Cezarina CHIVU, Dunarea de Jos University of Galati, Romania
Florin SUSAC, Dunarea de Jos University of Galati, Romania
Gabriel Radu FRUMUȘANU, Dunarea de Jos University of Galati, Romania
Mitică AFTENI, Dunarea de Jos University of Galati, Romania
Nicușor BAROIU, Dunarea de Jos University of Galati, Romania
Viorel PĂUNOIU, Dunarea de Jos University of Galati, Romania
Virgil Gabriel TEODOR, Dunarea de Jos University of Galati, Romania

Executive Chair
Marius Andrei MIHALACHE, Gheorghe Asachi Technical University of Iasi, Romania

Innovative Manufacturing Engineering and Energy - IMANEE2025   Materials Research Forum LLC
Materials Research Proceedings 61 (2026) 1-8   https://doi.org/10.21741/9781644903995-1

Experimental research on machining 2043 AL-LI alloy used in aircraft construction using uncoated and DLC-coated milling tools

Nicolae Ioan PASCA[1,a*], Mihai BANICA[1,b], Lucian BUTNAR[1,c],
Marius COSMA[1,d], Vasile NASUI[1,e]

[1]Technical University of Cluj-Napoca, North University Centre of Baia Mare, 62A, V. Babes Street, Romania

[a]nicu.pasca2604@gmail.com, [b]Mihai.Banica@imtech.utcluj.ro, [c]Lucian.Butnar@imtech.utcluj.ro, [d]Marius.Cosma@imtech.utcluj.ro, [e]Vasile.Nasui@imtech.utcluj.ro

Keywords: Alloy, Coating, Diamond, Machining, Tool

Abstract. The paper's objective is to show the durability obtained at the milling of Al-Li 2043 alloy, used in the aerospace industry. The study compares the behavior of cutters with uncoated inserts with those coated with DLC (Diamond-Like Carbon), evaluating parameters such as insert wear, surface roughness of the machined parts and insert durability. Experiments were performed under various cutting conditions to determine the influence of DLC coating on the process efficiency. The results revealed that DLC-coated cutters offer significant advantages, including longer tool life, making them a promising choice for precise machining of this advanced alloy. The study contributes to the optimization of manufacturing processes in the aeronautical field, providing solutions to improve productivity and quality of finished parts.

1. INTRODUCTION
1.1 The experimental methodology
The essential elements of the experimental methodology are:
1. Definition of the research problem: determining the durability of worn cutting inserts before they deteriorate and break during machining.
2. Hypotheses: DLC coating provides greater durability compared to uncoated inserts. Aggressive cutting parameters provide increased productivity but involve greater wear.
3. Variables:
   - independent variables: 3 milling parameters; 2 types of milling inserts.
   - dependent variables: durability, roughness, CNC spindle loads.
   - control variables: CNC machine, type of workpiece.
4. Experimental design:
   - experimental groups: group of uncoated inserts; group of DLC-coated inserts.
   - randomization of subjects: elimination of absurd results.
5. Data collection: standard form to track each set of uncoated/coated inserts.
6. Analysis of the data obtained: Excel files for data modeling.
7. Interpretation of the results: formulating conclusions based on the data obtained [1].

1.2 Research objectives
- Evaluation of the performance of DLC-coated milling cutters in machining Al-Li 2043 alloy, analyzing factors such as tool wear, surface roughness and process productivity.
- Optimization of machining parameters to obtain an ideal combination of surface quality and process efficiency.
- Identification of the interaction mechanisms between the DLC layer and the Al-Li alloy, providing insights into their tribological behavior.

Innovative Manufacturing Engineering and Energy - IMANEE2025
Materials Research Proceedings 61 (2026) 1-8

Materials Research Forum LLC
https://doi.org/10.21741/9781644903995-1

- Comparison of experimental results with the performance of uncoated or coated tools with other materials, to demonstrate the advantages of using DLC.

These objectives contribute to the development of innovative solutions for the aerospace industry, aiming at both improving the performance of machining tools and the efficiency of the manufacturing process.

2. STUDIED MATERIAL: 2043 T83

Aluminum-lithium alloys are a class of advanced materials widely used in the aerospace industry due to their unique combination of mechanical properties and low weight. The introduction of lithium into the composition of aluminum alloys provides a significant reduction in the material's density, thereby improving the strength-to-weight ratio [2].

Al-Li alloys, such as 2043, offer the following advantages:
- Superior mechanical strength: these alloys exhibit excellent tensile and fatigue strength, making them suitable for critical structural components.
- Corrosion resistance: the addition of lithium contributes to improved corrosion resistance, reducing the risk of damage in aggressive environments.
- Damping behavior: Al-Li alloys exhibit an increased ability to reduce vibrations, which improves safety and comfort during flight [3].

Aluminum alloys are used in the aeronautical industry depending on their properties. The main series of alloys used in the aeronautical industry are presented in Table 1.

Table 1. Main series of aluminum alloys [4]

Series	Main alloying element	Alloys used in the aeronautical industry
1xxx	99.0% Aluminium	
2xxx	Copper	2014, 2024, 2043, 2099, 2196, 2219, 2224
3xxx	Manganese	
4xxx	Silicon	
5xxx	Magnesium	5083, 5086
6xxx	Silicon and Magnesium	6063, 6061, 6082, 6092
7xxx	Zinc	7005, 7049, 7050, 7075, 7136, 7175, 7249
8xxx	Other elements	

When machining difficult materials such as Al-Li alloys, DLC coatings offer significant benefits:
- Wear resistance: DLC protects tools from wear caused by interaction with abrasive materials.
- Low friction coefficient: this facilitates chip evacuation and reduces heat generation, keeping the tool temperature within safe limits.
- Low adhesion: materials such as Al-Li tend to adhere to the surface of tools, but the DLC coating minimizes this phenomenon, extending the durability.
- Superior surface quality: due to the reduced friction, the surfaces obtained are smoother [5].

The **temper** is designated by a system of standardized codes defined by the Aluminum Association, and it refers to the state of material, specifically its mechanical properties and the thermal or mechanical treatments it has undergone. T grades for heat treatable alloys can have from one to five digits. The first digit after the T always indicates the basic type of treatment, and the second through fifth, if used, indicates whether the product has been stress relieved and, if so, how it was stress relieved and whether there have been any other special treatments.

The first digit after the T can be any of the following:
- T1: the alloy has been cooled directly after hot processes then naturally aged to a stable state.

- T2: the alloy has been cooled from a high temperature also after hot working processes and is then cooled before being naturally aged to a stable state.
- T3: the alloy has received a solution heat treatment after hot working, quenching, cold working and is naturally aged to a stable state.
- T4: the alloy has received a solution heat treatment and has been naturally aged to a stable state.
- T5: the alloy has been cooled after extrusion and is artificially cooled.
- T6: the alloy has been heat treated and artificially aged without significant artificial cooling.
- T7: the alloy has been solution heat treated and, without any significant cold working, is aged in a furnace to a stable state.
- T8: the alloy has been solution heat treated, cold worked to harden and then artificially aged to achieve precipitation hardening. A common example of an alloy that may have the T83 temper is 2024-T83, used in aerospace and other applications where high fatigue strength and good mechanical strength are required.
- T9: the alloy has been solution heat treated, artificially aged to achieve precipitation hardening, and then cold worked to improve its strength.
- T10: the alloy has been cooled after a high temperature machining process such as extrusion, cold worked, and then artificially aged to achieve precipitation hardening [6].

According to the customer's requirements, the parts on which the cutting tools presented in the experiment are tested will have to be machined from extruded semi-finished products made of alloy 2043 T83. The chemical composition of 2043 is presented in table 2.

Table 2 Chemical composition of alloy 2043 [7]

	Chemical composition											
	Cu	Li	Zn	Mg	Mn	Zr	Ti	Fe	Si	Ag	Others (Each)	Others (Total)
Min	2.5%	1.4%	-	0.1%	-	0.05%	-	-	-	-		
Max	3.1%	1.9%	0.3%	0.6%	0.3%	0.15%	0.1%	0.1%	0.08%	0.1%	0.05%	0.15%

3. RESOURCES NEEDED
3.1 The CNC machine and the clamping device
To carry out the experiments was used the Bavius PBZ HD 600 numerically controlled machine center which is shown in Figure 1. Of the 6 CNC's present in the *Mechanical Manufacturing Department* at *Universal Alloy Corporation Europe* Factory, we chose HD 600_2 for testing the cutting inserts.

The machining fixture used is machined from alloy 5083 and is used to machine several parts in the Airbus A350 aircraft beam family. The fixture is machined from a single piece, has channels according to the shape of the semi-finished product and channels for the vacuum installation so that the semi-finished product is held on the fixture after the removal of the screws and fixing clamps. The machining fixture is presented in Figure 1 and is dark grey in color. The yellow fixture represents the workpiece.

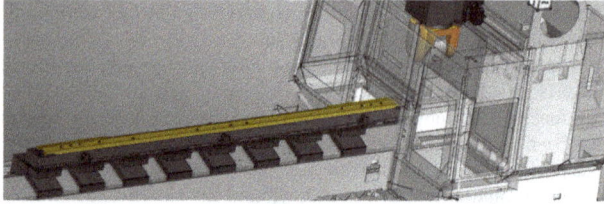

Fig. 1 The Machine-tool-clamping device-workpiece system

Bavius PBZ HD 600 technical data:
- Maximum dimensions of the workpiece: L = 12000 mm, l = 800 mm, h = 575 mm.
- Spindle power: 63-80 kW/h, depending on the model.
- Spindle maximum rotation speed: S = 30000 rpm.
- Maximum torque: 36 N/m.

3.2 The milling tools tested

The tool holder (Fig. 2 left) used has HSK 63-A grip. The model has 3 slots for mounting the cutting inserts. The cutting inserts have two corners that meet the workpiece and are presented in Figure 2 right. After the first corner has worn out, the insert is rotated for machining with the second corner. The inserts are fixed to the body of the milling cutter using fixing screws. [8]

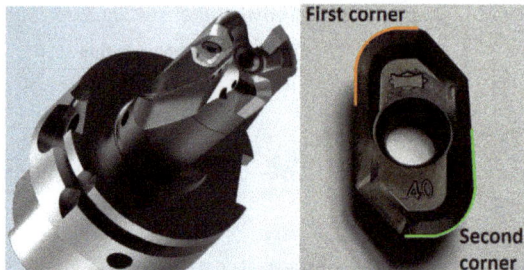

Fig. 2 Tool and insert [9]

3.3 Vibration measuring instrument

To measure the spindle's vibrations, we used Hofmann MI 2100 device. After each batch of tested inserts, we ran a diagnosis. The procedure is to increase the spindle's speed progressively by every 1000 rpm and maintain the speed for one minute for each speed.

3.4 Measuring devices

For roughness measuring Mitutoyo SJ-210 device was used because it is very useful because the device can work independently of the main supply.

Because the parts are machined in serial machining conditions, we need the measuring and control devices that are required in the inspection plans related to the work orders being processed. Therefore, we need a caliper, a caliper for measuring thickness, internal and external micrometer, GO/NOGO gauges, radius gauges and roughness meter.

Innovative Manufacturing Engineering and Energy - IMANEE2025   Materials Research Forum LLC
Materials Research Proceedings 61 (2026) 1-8   https://doi.org/10.21741/9781644903995-1

3. RESULTS AND DISCUSSION

The experiment was structured in 3 parts, each part using different cutting parameters.

I.   First test use: $S = 28000$ [rpm], $V_f = 14000$ [mm/min], $f_z = 0.167$ [mm/rev];
II. Second test use: $S = 28000$ [rpm], $V_f = 19000$ [mm/min], $f_z = 0.226$ [mm/rev];
III. Third test use: $S = 27000$ [rpm], $V_f = 16200$ [mm/min], $f_z = 0.2$ [mm/rev].

S – spindle speed; V_f – feed rate; f_z – feed per tooth.

The working parameters were chosen in such a way as not to cause damage to the CNC machine. After finding that the first combination of parameters was loose, we decided to increase spindle speed and the feed rate. After the second combination of parameters we decided to decrease spindle speed and calculate feed rate so that match the feed to 0.2 mm/rev/tooth.

The experiment is carried out in serial machining conditions. To operate the CNC machine, it requires one operator per shift, three shifts are planned for each day of each month, meaning continuous production. To carry out the experiment, 20 sets of inserts were used for each cutting parameter tested.

In total, the following were tested:

20 UC sets * 3 insert/set * 3 different parameters = 180 uncoated inserts (same batch)

20 DLC coated sets * 3 insert/set * 3 different parameters = 180 DLC coated inserts (same batch)

Terminology:

- 121-180 – represents the sets of cutting inserts;
- UC – symbolizes uncoated inserts / DLC – symbolizes Diamond-Like Carbon coated inserts;
- F1/F2 – represents the faces that will machine the workpiece;
- R_a – roughness (according with the customer requirements, maximum $R_a = 3.2$ μm).

A. Before starting the testing, we performed a spindle vibration diagnosis to see if the spindle would suffer after testing several cutting modes. The device used was the Hofmann MI 2000. The diagram is presented in Figure 3.

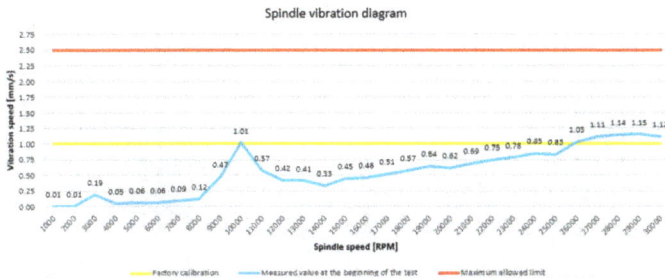

Fig. 3 Vibration diagnostics before testing

B. Testing uncoated inserts using $f_z=0.167$ mm/rev. Spindle loading did not exceed 15%. The vibration diagram remained unchanged. Average durability was 106 minutes. The roughness (R_a) was $0.866 - 1.127$ μm.

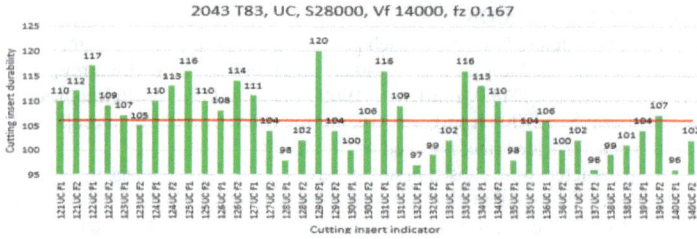

Fig. 4 Uncoated insert durability, working parameter f_z = 0.167 mm/rev

C. Testing uncoated inserts using f_z=0.226 mm/rev. Spindle loading reached 75%. The average durability was 92 minutes. R_a = 1.067 – 1.794 μm.

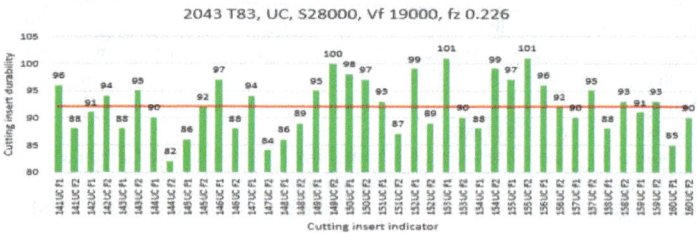

Fig. 5 Uncoated insert durability, working parameter f_z = 0.226 mm/rev

D. Testing uncoated inserts using f_z=0.2 mm/rev. Spindle loading reached 25%. The average durability was 126 minutes. R_a = 0.717 – 1.663 μm.

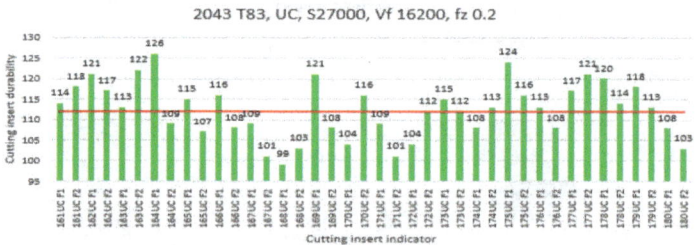

Fig. 6 Uncoated insert durability, working parameter f_z = 0.2 mm/rev

E. Testing DLC coated inserts using f_z=0.167 mm/rev. Spindle loading reached 60%. The average durability was 359 minutes. R_a = 0.862 – 1.591 μm.

Fig. 7 Diamond-Like Coated insert durability, working parameter $f_z = 0.167$ mm/min

F. Testing DLC coated inserts using f_z=0.226 mm/rev. Spindle loading reached 80%. The average durability was 257 minutes. $R_a = 1.173 - 1.905$ μm.

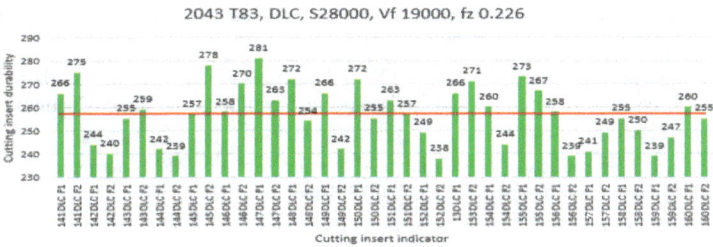

Fig. 8 Diamond-Like Coated insert durability, working parameter $f_z = 0.226$ mm/min

G. Testing DLC coated inserts using f_z=0.2 mm/rev. Spindle loading reached 60%. The average durability was 455 minutes. $R_a = 0.883 - 1.631$ μm.

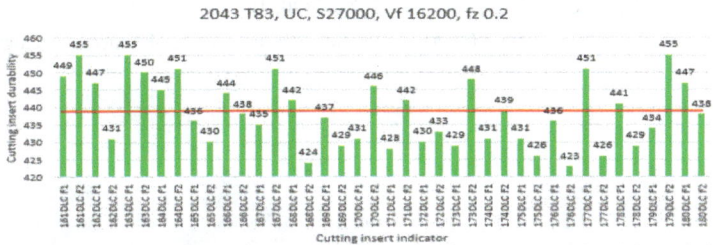

Fig. 9 Diamond-Like Coated insert durability, working parameter $f_z = 0.2$ mm/min

4. CONCLUSIONS

1. From the experimental research done, it is found that the durability of DLC coated cutting inserts is 2.8 to 3.9 times higher than the uncoated ones.

2. The parameter with the highest average minutes machined (439 minutes for DLC coated inserts) is $f_z = 0.2$ [mm/rev].

3. The most extreme parameter was $f_z = 0.226$ [mm/rev]. Here we had only 92 minutes machined for uncoated inserts and 257 minutes for DLC coated inserts. Also, higher values for spindle load and roughness were obtained.

4. The highest wear was for uncoated inserts using the parameter f_z=0.226 [mm/min]. 0.36 [mm] was measured.

5. The spindle vibration diagram underwent the greatest changes after machining with the parameter f_z=0.226 mm/min.

REFERENCES

[1] Kerlinger, F. N., Lee, H. B., Foundations of Behavioral Research, Holt, Rinehart, and Winston, New York, 2000

[2] Rioja, R.J., Liu, J., The Evolution of Al-Li Alloy Products for Aerospace and Space Applications, Metallurgical and Materials Transactions A, 43(9), 3325–3337, 2012. https://doi.org/10.1007/s11661-012-1155-z

[3] Davis, J.R., Aluminum and Aluminum Alloys, ASM International, ISBN 9780871704962, 1999. https://doi.org/10.31399/asm.tb.caaa.9781627082990

[4] Lumley, R.N., Introduction to aluminum metallurgy, Fundamentals of Aluminum Metallurgy (pp.1-19). https://doi.org/10.1533/9780857090256.1

[5] Wang, J., Komvopoulos, K., Friction and Wear Properties of Diamond-Like Carbon Coatings: Effects of Microstructure and Environmental Conditions, Surface & Coatings Technology, 326, 166–176, 2017

[6] Kaufman, J.G., Introduction to Aluminum Alloys and Tempers, ASM International, ISBN 0-87170-689-X, 2000, p39-76. https://doi.org/10.1361/iaat2000p039

[7] Certificate of Conformity, Universal Alloy Corporation, https://universalalloy.com/Certifications-NADCAP.html

[8] Pasca, N. I., Banica, M., Butnar, L., Nasui, V., Experimental research on machining with dlc diamond coated milling tools of aluminum alloy parts used in the aeronautical industry, Acta Technica Napocensis, Technical University of Cluj-Napoca, Series Applied Mathematics, Mechanics, and Engineering 2024. Vol. Vol 67, No 4 (2024)

[9] Mapal Microtek Catalog, NeoMill-Alu-QBig, https://mapal.com/en-int/media/catalogues

Innovative Manufacturing Engineering and Energy - IMANEE2025
Materials Research Proceedings 61 (2026) 9-16

Materials Research Forum LLC
https://doi.org/10.21741/9781644903995-2

Analysis of ISO 13485:2016 requirements and their effects on the medical device industry

Nicoleta-Lăcrămioara MANTA[1,a]* and Irina SEVERIN[1,b]

[1]National University of Science and Technology POLITEHNICA Bucharest, 313 Splaiul Independenței, 060042, Bucharest, Romania

[a]nicoleta.manta1602@stud.fiir.upb.ro, [b]irina.severin@upb.ro

Keywords: ISO 13485:2016 Requirements, Medical Device Industry, Corporate Performance Impact, Quality Engineering

Abstract. In this paper, the ISO 13485 standard was evaluated from multiple perspectives, including the reasons for its development, as well as its global applicability, particularly in Europe, the USA, and Canada. However, there is a lack of research regarding the implementation of the standard by leading companies in the medical device (MD) sector, which limits the understanding of its importance and impact. To address this gap, a case study methodology was employed, involving the identification of the top 10 companies in the MD sector, including Medtronic, Johnson & Johnson, Abbott, and Siemens Healthineers. For each company, aspects such as manufactured products, company size, and annual revenues were analyzed. The purpose of this comparative analysis was to determine whether these companies have implemented the ISO 13485:2016 standard and to assess the impact of this implementation on their performance, as well as to observe best practices and other standards implemented alongside it. The study results suggest that implementing ISO 13485:2016 can contribute to improving the performance and competitiveness of companies in the MD sector, facilitating access to international markets and ensuring compliance with industry-specific regulations. However, specific barriers encountered during the implementation process of the standard's requirements, changes in the number of non-conformities after implementation, or increases in the company's reputation were not specifically identified. These aspects require further research to thoroughly evaluate the impact of implementing this standard on various aspects of corporate performance. The current research was conducted based on publicly available online data. Future research needs to contact companies to obtain relevant information about best practices in implementing the standard, as well as about the quantifiable benefits brought about.

Introduction

Status of previous research

The reduction of barriers to the adoption of standards is a goal for those who are aware of the advantages of obtaining certification. Part of the efforts made to achieve this goal involves developing a set of educational materials, such as lesson plans, assignments, and content delivered in various forms. The creators of these materials aim to both contribute to the understanding of standards by students in the legal and professional context, as well as to "call to action" collaborators. One of the issues related to meeting standards is market access [1].

Regarding medical devices, the main stages of the life cycle of medical devices and associated services are regulated by the ISO 13485 standard. The first version of the standard was published in 1996, and the most recent update was released in 2016. One of the reasons the ISO 13485 standard was developed was to address the specific requirements of companies that manufacture medical and pharmaceutical devices. These special requirements are difficult to apply within the general ISO 9001 standard, which does not include regulations specific to the MD industry [2, 3].

A MD is a piece of equipment, instrument, apparatus, implant, or reagent used in vitro, designed to be used for the diagnosis, prevention, monitoring, or management of health issues [4].

ISO 13485 is a key standard used globally, including in Europe, the USA, and Canada, playing an essential role in the MD industry. In Europe, it is seen as a requirement for most devices, while in the USA, the FDA (Food and Drug Administration) aligns its regulations with it. The increasing importance of the standard, especially with emerging technologies like AI, makes it critical for stakeholders worldwide [5].

The regulations for medical devices in India are governed by the Medical Device Rule (MDR) 2017, which classifies devices based on risk levels and requires obtaining licenses from the Central Drugs Standard Control Organization (CDSCO), in compliance with ISO 13485:2016 for quality and safety assurance. In the EU, previous directives had deficiencies in post-market surveillance and clinical evaluation, which led to concerns regarding patient safety, especially in the context of the rise of software-based devices and personalized medicine. Therefore, the MDR introduced in 2021 was designed to align European regulations with international standards, improving safety, traceability, and post-market surveillance, while imposing strict requirements for clinical evaluation, risk classification, and compliance with ISO 13485:2016 for obtaining the CE mark.

In the USA, regulations are managed by the FDA through the Center for Devices and Radiological Health, which handles pre-market approval, inspections, and compliance with the ISO 13485:2016 standard. The 2016 law was adopted to accelerate innovation, simplify regulations, and ensure a balance between safety, efficacy, and rapid access to new technologies.

ISO 13485:2016 is a key global standard for MD compliance in the USA, EU, and other regions, facilitating market access and enhancing product safety, quality, and operational efficiency, while also providing a competitive advantage. The importance of global regulatory harmonization includes facilitating international trade by reducing barriers and promoting wider market access, stimulating innovation through investment and collaboration, and addressing global challenges such as technological advancements and public health emergencies [6].

The identified gaps refer to the lack of research conducted on top companies in the MD field, aimed at identifying the advantages brought by ISO 13485:2016 certification, as well as uncovering potential obstacles in obtaining this certification. This lack of research limits the understanding of both the complexity of the ISO 13485:2016 certification process and the importance of implementing this standard [7].

Paper structure
The following sections of the paper are structured as follows:
Section 2 presents the research methodology and the key aspects followed throughout the case study; Section 3 includes a case study that analyzes the top 10 companies in the medical industry, conducted in two stages. Stage 1 presents aspects such as the number of employees, the size of the organization, and the range of products made, while Stage 2 provides additional information about these companies, such as the certifications held and the certifying bodies that issued them. Section 4 highlights the conclusions, allowing the observation of the main findings, as well as the research gaps and future research directions.

Research methodology
The methodology used for this paper is the case study. It provides the opportunity to analyze top companies in the MD field, aiming to observe their main characteristics, such as size, product range, market reach, and market competition dynamics. Additionally, the case study allows for the investigation of the key certifying bodies for ISO 13485:2016, as well as the identification of other standards that are typically implemented alongside it, thus highlighting the complexity of the certification process. Another relevant aspect that can be drawn from the study is the time elapsed

since certification was obtained, suggesting that these companies did not reach the top immediately after certification but instead demonstrated continuous development.

Case study

The case study focuses on analyzing the top 10 companies in the MD field, including both essential financial data and relevant information about ISO 13485:2016 certification and other additional certifications obtained. Tables 1 and 2 provide a detailed overview of the characteristics of these companies, including turnover, number of employees, areas of activity, ISO certifications, and the certifying bodies involved.

Thus, TÜV (Technischer Überwachungsverein), DQS (Deutsche Gesellschaft zur Zertifizierung von Managementsystemen), SGS (Société Générale de Surveillance), UL (Underwriters Laboratories), Intertek, SUD (Société Unitaire de Développement), TÜV Rheinland and BSI (British Standards Institution) are international certification bodies, each with specific areas of expertise. They offer services such as evaluating and certifying product compliance with safety and quality standards, assessing quality, environmental, and information security management systems, as well as testing product safety and performance, particularly in the technological and household appliance sectors. Additional details about these bodies and their certifications are presented in Table 2, which provides a detailed comparison of the ISO 13485:2016 certifications and other certifications obtained by the top 10 companies.

Table 1 presents the top 10 global companies in the MD industry [8, 9, 10] as well as key indicators that define these organizations. Thus, one can see not only the individual performance of each company but also the main competitors in the market, considering that each company in the top 10 competes directly with the others. The top companies in the medical devices field offer a wide range of products and operate in over 100 countries, having a significant global impact. They are leaders in technological innovations and have substantial resources for continuous development. Large companies typically generate billions in revenue and have a considerable workforce, which allows them to meet global market demands and comply with the requirements of certifying bodies. The following table also presents data on the revenue evolution of companies between 2020 and 2024.

Table 1: Top 10 Global MD Companies and Key Indicators

No.	Company	Revenue [11]		No. of emplo-yees	Com-pany Size	Product Range	Coun-tries Sold In	Src.
		Evol. 2020-2024 [%]	2024 ($ B)					
1.	Medtronic	11,87	32,364	101,000	Large	Cardiovascular, neurovascular, orthopedic, diabetes, and other MDs	Over 140 ctrs.	[12]
2.	Johnson & Johnson	7,49	88,821	103,000	Large	Surgical, orthopedic, ophthalmic, cardiovascular, and other MDs	Over 175 ctrs.	[13]
3.	Siemens Healthineers	54,74 [14]	22.363 [14]	54,000	Large	Medical imaging, molecular diagnostics, and healthcare IT solutions	Over 70 ctrs.	[15]
4.	Philips Healthcare	-12,87	19,501	73,000	Large	Medical imaging, patient monitoring, and connected health solutions	Over 100 ctrs.	[16]
5.	GE Healthcare	14,67	19,672	56,000	Large	Medical imaging, patient monitoring, and healthcare IT solutions	Over 100 ctrs.	[17]
6.	Stryker Corporation	57,57	22,595	40,000	Large	Orthopedic, neurovascular, and endoscopic MDs	Over 100 ctrs.	[18]
7.	Abbott Laboratories	21,16	41,950	103,000	Large	Cardiovascular, diabetes, diagnostics, and other MDs	Over 150 ctrs.	[19]
8.	Boston Scientific	68,87	16,747	36,000	Large	Cardiovascular, urology, endoscopy, and other MDs	Over 100 ctrs.	[20]
9.	Zimmer Biomet	25,17	7,679	18,000	Large	Orthopedic, dental, and cranio-maxillofacial MDs	Over 100 ctrs.	[21]
10.	Becton Dickinson	25,34	20,178	70,000	Large	MDs for medication administration, diagnostics, and other applications	Over 150 ctrs.	[22]

Table 2 presents a comparative analysis of the 10 companies, focusing on ISO 13485:2016 certification, including information about the certifying bodies, as well as other additional certifications held by these companies.

Table 2: Certification Analysis & Certifying Bodies for the Top 10 MD Companies

No.	Company	Year of transition to ISO 13485	ISO 13485:2016 Certification	Certifying Body	Additional Certifica-tions	Technological innovations [23]
1.	Medtronic	1993	Yes	TÜV SÜD	ISO 9001:2015; ISO 14001: 2015	Launched the Evolut™ FX+ TAVR system, offering minimally invasive options for heart valve treatments.
2.	Johnson & Johnson	1996	Yes	SGS	ISO 9001:2015; ISO 14001: 2015	Continuously invests in technology to launch new products and improve patient outcomes.
3.	Siemens Healthineers	1995	Yes	DQS	ISO 9001:2015; ISO 14001: 2015	Known for advanced medical imaging and molecular diagnostic solutions.
4.	Philips Healthcare	1997	Yes	BSI	ISO 9001:2015; ISO 14001: 2015	Offers medical imaging and patient monitoring solutions, enhancing patient care.
5.	GE Healthcare	1998	Yes	UL	ISO 9001:2015; ISO 14001: 2015	Develops advanced imaging and monitoring solutions to improve diagnosis and treatment.
6.	Stryker Corporation	1999	Yes	Intertek	ISO 9001:2015; ISO 14001: 2015	A leader in orthopedic and neurovascular devices, improving patient outcomes through innovation.
7.	Abbott Laboratories	2000	Yes	TÜV Rheinland	ISO 9001:2015; ISO 14001: 2015	Develops cardiovascular devices and diagnostic solutions to advance patient care.
8.	Boston Scientific	2001	Yes	BSI	ISO 9001:2015; ISO 14001: 2015	Develops minimally invasive devices for cardiovascular and urological treatments.
9.	Zimmer Biomet	2002	Yes	TÜV SÜD	ISO 9001:2015; ISO 14001: 2015	Creates orthopedic devices to improve mobility and functionality.
10.	Becton Dickinson	2003	Yes	UL	ISO 9001:2015; ISO 14001: 2015	Develops solutions for medication delivery and diagnostics, significantly impacting patient care.

Based on the presented data, it is observed that all 10 top companies in the medical devices field are certified ISO 13485:2016, highlighting the importance of this certification for the quality and compliance of products. These companies are also additionally certified according to ISO 9001 and ISO 14001 standards, validating their processes through certification bodies such as TÜV SÜD, BSI, and UL. Furthermore, the companies have demonstrated a continuous commitment to technological innovations, launching advanced solutions in cardiovascular treatments, medical imaging, and orthopedic devices, which have had a significant impact on improving patient outcomes and the quality of medical care.

Conclusions

Main conclusions

In conclusion, ISO 13485:2016 certification is an indicator of companies' commitment to quality and safety, being a determining factor in their success and reputation in the global MD market. These are the effects of the ISO 13485:2016 standard on the MD industry, highlighting the importance of certification in ensuring the quality and compliance of products.

The top companies in the medical devices field offer a wide range of products, covering areas such as cardiology, orthopedics, medical imaging, and diagnostics. They are active in over 100 countries, demonstrating a significant global impact. These companies are leaders in developing advanced technological solutions, with a large number of employees and consistent revenues. They continue to influence the industry's evolution through investments in research and development. However, the degree of impact of ISO 13485:2016 certification varies between companies, depending on factors such as company size and the products manufactured. For instance, companies like Johnson & Johnson, with a larger workforce and broader product portfolio, have a more extensive global presence compared to Siemens Healthineers, which operates in over 70 countries [14,15].

All these firms are major players in the global market, with billions in revenue and a significant workforce. This allows them to expand their operations and meet global market demands. Furthermore, some companies, like Medtronic, with revenues exceeding $32 billion, show a larger financial capacity, which may allow for more extensive investments in technological innovations compared to companies like Zimmer Biomet, with a significantly smaller revenue of $7.7 billion [13, 21]. Additionally, while companies like Abbott Laboratories, with a broad product range including diagnostics and cardiovascular devices, continue to push the boundaries of patient care, others like Zimmer Biomet, focusing predominantly on orthopedic devices, may have a more specialized impact [19, 21].

Based on the presented data, it is observed that all 10 top companies in the medical devices field are certified ISO 13485:2016, highlighting the importance of this certification for the quality and compliance of products. These companies are also additionally certified according to ISO 9001 and ISO 14001 standards, validating their processes through certification bodies such as TÜV SÜD, BSI, and UL. Furthermore, the companies have demonstrated a continuous commitment to technological innovations, launching advanced solutions in cardiovascular treatments, medical imaging, and orthopedic devices, which have had a significant impact on improving patient outcomes and the quality of medical care.

Regarding revenue evolution between 2020 and 2024, Boston Scientific recorded the highest growth rate, at 68.87%, followed by Stryker Corporation with 57.57% and Siemens Healthineers with 54.74%. The only company to experience a decrease in revenue during this period was Philips Healthcare, which saw a reduction of 12.87%. Through the analysis of the annual sales evolution, it can be concluded that the implementation of QMS appears to have significantly contributed to revenue growth [11, 14].

Limitations and future research directions

The data identified for the 10 companies presented is limited and does not allow for the identification of the challenges encountered during the implementation of the ISO 13485:2016 standard or the actual contribution of the standard by providing the opportunity to compare data such as changes in the number of non-conformities, customer satisfaction evolution, or new markets accessed.

An important aspect for future research will be contacting the companies to obtain relevant information regarding best practices in implementing the standard, as well as to assess the quantifiable benefits.

References

[1] M. Williams, D. Solomon, C. Drummond, Moot court cases: Bringing standards to life, 2024. https://doi.org/10.18260/1-2-39254

[2] International Organization for Standardization, DIN EN ISO 13485:2016 – Medical devices – Quality management systems – Requirements for regulatory purposes, ISO, Geneva, 2016.

[3] NQA, Full guide to ISO 13485 – medical devices. https://www.nqa.com/en-us/resources/blog/february-2017/a-guide-to-iso-13485

[4] International Organization for Standardization, ISO 13485: Medical devices — Quality management systems – Requirements for regulatory purposes, ISO, Geneva, 2016. https://www.iso.org/iso-13485-medical-devices.html

[5] I. Juuso, Developing an ISO 13485-certified quality management system: An implementation guide for the medical-device industry, Routledge, 2022. https://doi.org/10.4324/9781003202868

[6] N. M. Jadhav and R. S. Shendge, ISO 13485:2016 – The gateway of global or regional harmonization for medical device regulations, International Journal of Pharmaceutical Quality Assurance 15 (1) (2024) 502–511. https://doi.org/10.25258/IJPQA.15.1.76

[7] W. Linck, M. A. C. Tinoco, S. V. Bonato, I. H. Grochau, D. A. de J. Pacheco, C. S. Ten Caten, A maturity assessment methodology for ISO 13485 implementation in the medical devices industry, International Journal of Quality and Reliability Management. (2024). https://doi.org/ 10.1108/IJQRM-05-2024-0163.

[8] The top medical device companies of 2023. https://www.mddionline.com/business/top-40-medical-device-companies

[9] Proclinical, Top 10 medical device companies 2024. https://www.proclinical.com/blogs/2024-10/top-10-medical-device-companies-in-the-world-in-2024

[10] Medical Device Directory, Top ISO 13485 consultants – Compare 64 companies. https://www.medicaldevicedirectory.com/category/regulatory-quality-control/iso-13485-consultants

[11] Macrotrends – The long term perspective on markets. https://www.macrotrends.net

[12] Medtronic – Homepage. https://www.medtronic.com/ro-ro/index.html

[13] Johnson & Johnson – Changing health for humanity. https://www.jnj.com/

[14] Siemens Healthineers AG – Stock data. https://www.boerse-frankfurt.de/aktie/siemens-healthineers-ag/kennzahlen

[15] Siemens Healthineers – Corporate homepage. https://www.siemens-healthineers.com/

[16] Philips Healthcare – Innovative medical solutions. https://www.usa.philips.com/healthcare

[17] GE HealthCare - Medical systems and solutions. https://www.gehealthcare.com/

[18] Stryker – International homepage. https://www.stryker.com/

[19] Abbott – Life-changing health care technologies. https://www.abbott.com/

[20] Boston Scientific – Advancing science for life. https://www.bostonscientific.com/en-US/home.html

[21] Zimmer Biomet – Innovations in orthopedic care. https://www.zimmerbiomet.com/en

[22] Philips – About us. https://www.usa.philips.com/a-w/about

[23] BioSpectrum Asia, MedTech giants drive innovation and growth: Key developments and achievements. https://www.biospectrumasia.com/analysis/29/24567/medtech-giants-drive-innovation-and-growth-key-developments-and-achievements.html

Innovative Manufacturing Engineering and Energy - IMANEE2025   Materials Research Forum LLC
Materials Research Proceedings 61 (2026) 17-27   https://doi.org/10.21741/9781644903995-3

From hobby to 3D printing producer: A multi-perspective analysis of entrepreneurial evolution in Romania

Florin Ştefan JINGA[1,a], Irina SEVERIN[1,b]

[1]Splaiul Independenţei nr. 313, sector 6, Bucharest, Romania, Faculty of Industrial Engineering and Robotics (FIIR), National University of Science and Technology POLITEHNICA Bucharest

[a]florin_stefan.jinga@stud.fiir.upb.ro, [b]irina.severin@upb.ro

Keywords: FDM 3D Printing, Small-Scale 3D Printing, Hobby 3D Printing, 3D Printing Entrepreneurship, Limited Resources Business, Romanian 3D Printing

Abstract. The evolution of Fused Deposition Modeling 3D printing has encouraged hobbyists to transition into entrepreneurs, creating small-scale businesses in resource-limited environments. This study investigates Romanian 3D printing producers, analyzing their activity models, technical capabilities, challenges, achievements and market positioning through survey data and interviews. Findings reveal that most producers operate as self-taught professionals, leveraging affordable printers (BambuLab, Creality, Prusa), PLA/ABS/PETG materials and social media marketing. Key challenges include unrealistic customer expectations, market saturation and financial barriers, yet some overcome these through niche specialization and hybrid manufacturing approaches. This research highlights factors influencing business sustainability, offering insights into growth potential and resilience of Romania's decentralized 3D printing sector.

Abbreviations and definitions
AM - additive manufacturing
FDM - fused deposition modeling
PLA - polylactic acid
ABS - acrylonitrile butadiene styrene
ASA - acrylonitrile styrene acrylate
PETG - polyethylene terephthalate glycol
PC - polycarbonate
DIY - do it yourself
CNC - computer numerical control
OLX - online sales platform
SEO - search engine optimization
Small-scale 3D printing is used in this article to define the landscape of producers using budget desktop 3D printers, working in limited-resource environments without being part of large businesses, without benefiting of investment budgets.

Introduction
Context and motivation
This paper intends to explore the FDM (Fused Deposition Modeling) 3D printing entrepreneurial landscape in Romania, where technological advancements have enabled individuals to adopt a hobby and transition to service providers.[1][2] As additive manufacturing evolves, desktop 3D printers are becoming more and more accessible, allowing passionate individuals to enter the manufacturing sector with minimal investment.[3][4]

As the popularity grows it raises the question: "How do FDM 3D printing entrepreneurs operate, and what factors influence their business models, production processes, and market interactions?"

This study seeks to understand the profile of 3D printing entrepreneurs, their business models, manufacturing processes, client interactions, challenges, achievements and future perspectives. By analyzing their experiences, this research aims to support small-budget entrepreneurial engineering initiatives and provide insights into the sustainability of independent manufacturing ventures.[5][6][7]

Research problem and objectives
Despite the increasing adoption of FDM 3D printing as a manufacturing tool, little research has been conducted on small-scale producers operating in resource-limited environments.[2][8] Existing studies primarily focus on industrial-scale additive manufacturing, overlooking the localized, decentralized production networks formed by independent entrepreneurs.[1][5][9]

To fill this gap, the study aims to:
O1: Understand the profile of the producer.[5][10][11]
O2: Identify the motivations behind starting a small-scale 3D printing production.[12][13][14]
O3: Analyze the business models and how these entrepreneurs structure their operations.[1][9][15]
O4: Examine the challenges encountered in activity and achievements accomplished throughout time.[16][17]

By achieving these objectives, the study contributes to the broader understanding of digital manufacturing and decentralized production models, offering insights for future entrepreneurs and policymakers.[3][4][7][18]

Methodological approach
This research employs a triangulated methodology that combines:[19]
Survey – a structured questionnaire targeting 3D printing entrepreneurs in Romania
Semi-structured interviews – in-depth discussions with selected producers to gain qualitative insights into their operations
Literature review – analyzing existing research on small-scale 3D printing, business models, and digital manufacturing ecosystems.

Literature review
Understanding the dynamics of small-scale FDM 3D printing producers require a multi-theoretical perspective, integrating innovation adoption, entrepreneurial behavior and production decentralization. The literature reflects a convergence of user innovation theory, technology acceptance models and evolving business strategies within additive manufacturing.

One of the core theoretical pillars supporting this study is the concept of *user-driven innovation and prosumerism*, where individuals transition from passive consumers to active creators. Halassi, S., Semeijn, J. and Kiratli, N. explain that this phenomenon is facilitated by the accessibility of FDM technology, allowing users to not only fabricate products for personal use but also commercialize their creations through niche offerings and service-based models.[5] Rayna and Striukova highlight this evolution as a gradual progression from prototyping to home fabrication, illustrating how accessible design tools and decentralized production encourage the ascension of small-scale entrepreneurs.[1]

The *Technology Acceptance Model (TAM)* has also been widely used to explain adoption behavior in the early stages of 3D printing context. According to Schniederjans, early adopters exhibit high levels of self-efficacy and technical curiosity.[10]

Motivation for adopting 3D printing in entrepreneurial contexts has also been explored by researchers. Villanueva Orbaiz and Arce-Urriza identify intrinsic motivators such as enjoyment, learning and creative expression, as well as practical motivators like use-value and task efficiency. Sustainability has emerged as another motivational driver, entrepreneurs getting involved into

Innovative Manufacturing Engineering and Energy - IMANEE2025   Materials Research Forum LLC
Materials Research Proceedings 61 (2026) 17-27   https://doi.org/10.21741/9781644903995-3

circular economy practices, such as recycling plastic waste into filaments or promoting durable product alternatives.[12]

From business strategy perspective, the evolution of AM business models is reflected in the transition from simple prototyping to end-to-end service delivery, including consulting, digital design and custom fabrication. Chen's research on service-oriented manufacturing shows how small producers leverage digital tools and online platforms to offer tailored services, effectively bypassing traditional manufacturing limitations.[15] Decentralized and localized manufacturing has also been explored extensively by Ben-Ner and Siemsen, who argue that additive manufacturing empowers small-scale businesses by lowering the market entry barriers and reducing supply chain dependencies.[9]

However, despite these opportunities, barriers to adoption and scalability persist. As noted by Nie J., Xu X., Yue X., Guo Q. and Zhou, Y. small producers often face limitations in production speed, material quality and machine durability, especially when operating with constrained resources. These challenges refer to financial risks and customer related difficulties, as reported in recent studies comparing SME and industrial-scale AM adoption strategies.[16]

Finally, future-oriented studies such as "Predicting the Future of Additive Manufacturing" forecasting the consolidation of digital marketplaces and decentralized production networks as dominant trends in additive manufacturing.[18]

These insights support the relevance of investigating the entrepreneurial ecosystem in Romania, where informal production, niche specialization and community-driven learning define the local AM landscape.

Research methodology

This study is using a triangulated research approach, integrating an online survey, a semi-structured interview and literature review to explore the entrepreneurial landscape of small-scale FDM 3D printing businesses in Romania. With this mixed-method approach, data reliability is enhanced, providing a comprehensive understanding of producer profile, business model, operational challenges and market dynamics.[19]

Research design

A mixed-method approach was used to quantify trends through survey data while gaining qualitative insights from focus group interviews. Findings were contextualized within existing literature to accomplish the objectives and validate the results, aligning with studies on FDM 3D printing entrepreneurial landscape.[19]

Data collection
Online survey

A structured questionnaire was shared into Romanian FDM 3D printing social media groups gathering insights on:

Demographics and activity profile – age, education, activity type (hobby, freelancing, registered company), and operational region.

Production and business operations – material and equipment used, service offered, pricing strategy and production scale.

Market engagement – customer base (B2B or B2C), sales channels and marketing strategies.

Challenges, achievements and growth perspectives – barriers encountered and financial and operational limitations.

Semi-structured interview
In depth interviews were conducted with a focus group of 3D printing entrepreneurs, covering:
Entrepreneurial motivation – transition from hobby to business, self-taught learning and professional influences.
Business model and strategies – how services are structured, customer interaction and pricing approaches.
Technical and market challenges – limitations related to equipment and material and issues coming from market interaction.
Achievements and future perspectives – remarkable projects, scalability, specialization and potential shifts in the Romanian 3D printing sector.

Literature review
A systematic review was conducted to link empirical findings with existing 3D printing scientific research:
Producer profile – aligning with theories on prosumerism[5] and technology adoption[10].
Motivations for 3D printing entrepreneurship – examining intrinsic motivation[12] and sustainability drivers[20]
Business model evolution – comparing Romanian businesses to service oriented 3D printing models[6] and decentralized manufacturing[9].
Challenges and scalability – investigating cost barriers[16] and sustainability challenges[17].

Data analysis
Quantitative analysis applied to survey data using descriptive statistics to identify patterns in business growth and correlations between business maturity, specialization and sustainability efforts.[19]

Qualitative analysis is used to extract common themes from interviews transcripts, such as transition from hobby to professional, key business strategies, perceived industry challenges and mitigation approaches.[19]

Comparative triangulation by cross referencing the findings from all three sources is helping to validate the answer consistency, identify gaps between industry best practices and entrepreneurial approaches and ensure a comprehensive understanding of Romanian 3D printing entrepreneurship.[19]

Empirical research
Survey results
This section presents the findings of the online survey conducted among small-scale FDM 3D printing producers in Romania. To collect data a Microsoft forms questionnaire has been shared into 3 active 3D printing online communities collecting 70 respondents that have answered the survey. The results reflect the demographic profile, business models, materials and equipment used, as well as marketing strategies, customer interaction, challenges and achievements. Each subsection highlights key questions, accompanied by visual representation of the responses to provide a concise and intuitive understanding of the data.

Materials Research Forum LLC
https://doi.org/10.21741/9781644903995-3

Table 1. Survey results

Age distribution

12%	6%	31%	31%	12%	6%
<18	18-24	25-34	35-44	45-54	55-64

Gender distribution

6%	94%
Female	Male

Education level

19%	31%	25%	19%	6%
High school	Bachelor's	Master's	PhD	Unspecified

Primary occupation

25%	75%
3D Printing Service Provider	Other

Producer region

27%	23%	14%	9%	7%	5%	5%	5%	5%
Transilvania	Muntenia	Moldova	Oltenia	Banat	Maramureş	Bucovina	Crişana	Dobrogea

Producer motivation

29%	39%	11%	18%	4%
Hobby turned into business	Passion for tech & innovation	Market gap opportunity	Additional income	Academic background

First contact with FDM 3D printing

50%	25%	19%	6%
Social Media	School/ University	Self-research	Workplace

Learning method

79%	21%
Self-taught	Specialized courses

Activity model

50%	50%
Self-owned business	Hobby

Team size

62%	19%	12%	6%
1 person	2 people	3 people	5 people

Services offered

28%	30%	19%	15%	9%
3D Printing Only	3D Design & Modeling	Prototyping	Small-Series Production	Print Farm (High Volume)

Number of printers used

27%	13%	13%	13%	7%	7%	20%
1	2	3	5	8	15	>15

Popular equipment brands

44%	20%	16%	8%	4%	4%	4%
Bambulab	Creality	Voron	Bcn3d	Prusa	Flsun	Elegoo

Frequently used materials

31%	31%	18%	11%	9%
PLA	PETG	Mechanical materials	TPU	ABS

Average material quantity used per month

44%		28%	11%	17%
1-5 kg		5-10 kg	10-50 kg	>50 kg

Average monthly orders

47%		20%	13%	20%
1-5		5-10	10-15	>15

Services marketing

42%		26%	21%	11%
Community engagement		Social media ads	Website & SEO	Word of mouth

Sales channels

31%	31%	27%	12%
Direct sales	Social media platforms	Online marketplaces	Local shops

Customers

68%	32%
Individual customers	Legal entities

Industries where clients are active

Automotive	Hospitality	Medical
Advertising	Education	Carpentry
Agriculture	Household	Spare parts

Orders flow

Initial contact → Requirements and clarification → Offer and price negotiation → Approval and production → Delivery

Common challenges

Legal and fiscal pressure.
Market limitations.
Technical and resource constraints.
Intellectual property and sales.

Simplest products manufactured	Complex products manufactured
Functional everyday items (washers, cable holders, hooks for tools, spare parts for households). Small decorations or toy prints (keychains, thematic bookmarks). Technical or prototype components (test specimens, corner brackets, TPU gaskets, identification logos).	Technical (gearbox, hose guide with bearings, drone components, automotive parts, printed furniture). Artistic pieces (miniature figurines, large statues, decorative ornaments). Exhibition products (topographic map of Europe, scale model of a TAF (Romanian forestry tractor). Innovative and niche products (modular ant nest system, stamps for leather industry).

Achievements

Technical progress and projects (Equipment upgrades and improved productivity).
Personal and educational growth (Skill growth leads to better service offered).
Business endurance and recognition (In such a new economic landscape for Romania, entrepreneurs continue to emerge and exist).
Emotional impact (The success in the industry increases self-confidence and a sense of accomplishment).

Innovative Manufacturing Engineering and Energy - IMANEE2025   Materials Research Forum LLC
Materials Research Proceedings 61 (2026) 17-27   https://doi.org/10.21741/9781644903995-3

The survey reveals a diverse and resilient community of small-scale FDM 3D printing producers in Romania, shaped by creativity, adaptability a strong DIY mentality. While most operate with limited resources and small teams, the producers demonstrate remarkable versatility in product offerings, customer engagement and technical solutions.

Data concludes that despite facing challenges such as regulatory ambiguity and limited market awareness, many producers manage to overcome these barriers through continuous self-learning, driven by a strong sense of personal achievement and reinforcement by growing customer recognition.

Interview testimonials summary
Key similarities across the 3D printing producers
1. Motivation and entry into 3D printing

Most producers began 3D printing as a personal hobby, later evolving into business activities. Entry into the field was typically driven by curiosity or a specific project need. Nearly all were self-taught, relying on online resources and forums, often supported by prior experience in engineering, CNC machining or design.

2. Business models and services offered

Despite varying scales, producers commonly offer end-to-end services including consulting (identification of customer requirements), 2D and 3D design and 3D printing. Businesses are often run independently or in small teams and support both simple consumer requests and complex industrial prototypes. Personalization and adaptability are key to their service models.

3. Marketing and customer acquisition

Promotion is primarily done through OLX, Facebook and Instagram, with some investing in Google Ads and SEO. Word-of-mouth referrals are especially important for securing recurrent satisfied clients. Exposure strategies are delimited by budget limitations and market access.

4. Challenges faced

Producers face challenges such as unrealistic customer expectations, pricing pressures from unregulated competition, and limited resources. Technical limitations, especially in handling advanced materials or maintaining print quality, also restrict capabilities. Formalizing businesses is complicated by bureaucracy and inconsistent regulations.

5. Equipment and material preferences

BambuLab, Creality and Prusa printers are commonly used, selected for reliability and affordability. PLA and PETG dominate due to ease of use, while advanced users also work with ABS, ASA, Nylon and PC, particularly for functional industrial parts. Material choice is often constrained by equipment and customer cost sensitivity.

Key differences among the producers
1. Level of business maturity

Some producers (30%) operate formal business with high production volumes and industrial clients, while others remain freelancers (20%) or hobbyist (40%) testing market potential. A few (10%) have exited the market due to financial or operational difficulties.

2. Customer base

More advanced producers serve B2B clients in sectors like automotive, agriculture or manufacturing. Others cater mainly to individuals with custom or artistic requests. This segmentation influences material use, project complexity and pricing strategy.

3. Specialization areas

Specializations range from industrial prototyping and reverse engineering to artistic modeling and niche products like insect habitats. The diversity reflects the producer's technical background and market focus.

4. Production capacity and workflow adaptations
Monthly production varies widely, from a few kilograms to over 10 kg of filament. High-volume producers invest in printer enclosures, 3D scanners or custom setups, while smaller-scale operators rely on stock printers and freely available 3D models.
5. Future business outlook
Freelancers aim to scale and formalize operations, while more established businesses tend to stabilize or consolidate due to market saturation. Concerns include low-quality competition and a lack of differentiation among newer service providers.

Discussions and findings
This study investigated the entrepreneurial landscape of Romanian FDM 3D printing producers through a triangulated methodology, combining survey data, interview narratives and academic literature. The goal was to understand how low-resource producers operate, evolve and sustain their businesses in a developing additive manufacturing ecosystem.

Producer profiles and entry pathways
Romanian producers reflect a clear prosumer trend, often transitioning from passionate hobbyists to small-scale entrepreneurs. Both survey and interview data show that most participants are young (20-40 years old), well-educated and self-taught, aligning with global patterns of early technology adopters and user-driven innovation.[5][10][18] Entry into the field is frequently informal, motivated by creativity, personal interest or skill-building, rather than formal business intent.

Motivation and entrepreneurial drivers
The transition from hobbyist to entrepreneur is typically fueled by intrinsic motivation such as the desire to create, learn and experiment. Some producers were also driven by practical needs, such as replacing broken parts or exploring market gaps (e.g. ant habitats, niche art objects). This is also confirmed by other researchers identifying that producers are getting encouraged in their activity by intrinsic factors and practicality.[2][15][19] While sustainability was not a dominant theme in Romanian producer's narratives, a few interviewees and survey respondents demonstrate interest in recycling, material efficiency or offering durable alternatives.

Business models and operational strategies
The majority of Romanian producers operate solo or in small partnerships, offering a combination of 3D printing, 3D modeling (consulting, 2D and 3D design) and small-series production. Business models reflect service-oriented and platform-based approaches, where digital skills and customer interaction take precedence over mass production. Interviews confirmed a growing trend toward reverse engineering, CAD design and high-performance printing in more mature businesses. This evolution mirrors literature that tracks 3D printing firms moving from prototyping toward custom manufacturing and digital services.[1][9][15]

Technical capabilities and market engagement
The study highlights varied technical capacities, from single-printer setups to semi-industrial operations with 15-35 printers. Popular printers include BambuLab, Creality and Prusa, with materials such as PLA, PETG, ABS and specialized filaments (e.g. ASA, Nylon, carbon fiber infused) used depending on client needs. Market interaction is primarily digital, leveraging OLX, Facebook and Instagram, while communication happens through informal channels like WhatsApp. Both interview and survey data show that trust, responsiveness and flexibility are key for client retention.

Innovative Manufacturing Engineering and Energy - IMANEE2025   Materials Research Forum LLC
Materials Research Proceedings 61 (2026) 17-27   https://doi.org/10.21741/9781644903995-3

Challenges and constraints

Despite strong technical proficiency,

Despite technical proficiency, producers face numerous constraints like administrative and fiscal challenges discouraging formalization, customer misconceptions (e.g. expecting "photocopier-like" ease or ultra-low prices) complicating service delivery, unregulated competitors, with no quality control, using online model repositories undercutting prices, undermining quality-focused businesses and limited equipment capabilities or high material and operational costs further hinder scalability. This aligns with literature on scalability limits and resource bottlenecks faced by low-budget additive manufacturing ventures.[16][17]

Achievements and evolution

Despite obstacles, many producers have built loyal customer bases, created original product lines, produced functional parts for major industrial clients and achieved operational sustainability through self-learning, digital engagement and community trust. These micro-successes confirm theories of bottom-up innovation and agile entrepreneurship found in 3D printing literature.[8][18]

Reflections and contributions

Romania's small-scale FDM 3D printing producers often demonstrate similar starting points such as self-learning, hobbyist experimentation and early exposure to accessible desktop 3D printing equipment. While they share similar motivations and operational approaches, the community displays notable variation in terms of scale, specialization and market engagement. Some producers focus on B2C services such as customized household objects or artistic models, while others serve B2B clients in industrial, automotive or medical sectors. Growth tends to be driven by adaptability, technical proficiency and a strong DIY mentality, whereas stagnation is frequently linked to financial constraints, bureaucratic hurdles and the increasing pressure from informal competitors offering low-cost, low-quality services.

Despite their technical competence, many of these producers encounter persistent constraints that limit scalability and formalization. Administrative and fiscal barriers, such as complex registration procedures and inconsistent tax frameworks, discourage entrepreneurs from legitimizing their operations. Customer misconceptions regarding pricing, speed and the capabilities of 3D printing often create blockers during service delivery, while unregulated competitors who reuse online models and undercut prices erode trust in professional services. Technical bottlenecks, such as limitations in build volume, the absence of printer enclosures for advanced materials and elevated material and maintenance costs further restrict operational expansion. These issues align with the literature on resource bottlenecks and infrastructural limitations observed in low-budget additive manufacturing contexts.

At the same time, this study offers several original contributions to the academic discourse on 3D printing entrepreneurship. First, it provides a detailed, empirical exploration of Romanian producers. Second, it documents the transition from hobbyist to professional, highlighting the intersection between personal motivation, technical competence and entrepreneurial evolution. Moreover, the findings emphasize how local producers rely on informal networks, online communities and social platforms for marketing, learning and customer acquisition. In the absence of institutional or financial support, resilience is often sustained by a combination of personal satisfaction, skill development and increasing customer recognition. These insights enrich our understanding of how low-resource producers are not only navigating, but gradually shaping localized ecosystems of digital manufacturing.

Conclusion

This study explored the emerging entrepreneurial landscape of Romanian FDM 3D printing producers using a triangulated methodology built on survey data, interview testimonials and

Materials Research Forum LLC
https://doi.org/10.21741/9781644903995-3

literature analysis. It directly addressed 4 research objectives by examining the producer profile, entrepreneurial motivations, business structures and the challenges and achievements that define this segment. The results revealed that Romanian producers are typically young, self-taught and technologically inclined, reflecting the global profile of early adopters in additive manufacturing. Motivations to engage in 3D printing range from creative self-expression and technical exploration to market-driven need and supplementary income generation. As their businesses mature, many producers transition from casual services to complex project creation.

The study confirmed that business models vary significantly in scale and complexity, with operations spanning from single-person freelance services to semi-industrial print farms. Producers predominantly operate in a decentralized and resource-limited environment, using online sales platforms and social media to connect with clients while relying on personal networks to build trust and ensure continuity. Technical capabilities are often enhanced through DIY modifications and reverse engineering, especially among producers who specialize in industrial prototyping or functional components. Challenges such as fiscal ambiguity, equipment limitations and unregulated competition persist, yet many overcome these barriers through self-directed learning, practical innovation and commitment to customer service.

By highlighting these dynamics, the research contributes to a better understanding of the intersection between innovation, decentralized production and entrepreneurial resilience in emerging economies. The Romanian landscape highlights the potential of small-scale 3D printing as a sustainable and adaptable form of local manufacturing. However, to enhance this potential, supportive policies, access to scalable infrastructure and targeted educational programs will be essential. In doing so, these micro-entrepreneurs can transition from isolated initiatives into more integrated contributors to digital regional manufacturing.

References

[1] Rayna, T., Striukova, L., From rapid prototyping to home fabrication: How 3D printing is changing business model innovation, Technological Forecasting & Social Change, ISSN 0040-1625, Elsevier Inc., 2016. https://doi.org/10.1016/j.techfore.2015.07.023

[2] Wang, Q., Sun, X., Cobb, S.V.G., Lawson, G., Sharples, S., 3D Printing System: An Innovation for Small-Scale Manufacturing in Home Settings - Early Adopters of 3D Printing Systems in China, International Journal of Production Research, ISSN 0020-7543, Taylor & Francis, 2016. https://doi.org/10.1080/00207543.2016.1154211

[3] Attaran, M., Additive Manufacturing: The Most Promising Technology to Alter the Supply Chain and Logistics, Journal of Service Science and Management, ISSN 1940-9893, Scientific Research Publishing, 2017. https://doi.org/10.4236/jssm.2017.103017

[4] Weller, C., Kleer, R., Piller, F.T., Economic Implications of 3D Printing: Market Structure Models in Light of Additive Manufacturing Revisited, International Journal of Production Economics, ISSN 0925-5273, Elsevier, 2015. https://doi.org/10.1016/j.ijpe.2015.02.020

[5] Halassi, S., Semeijn, J., Kiratli, N., From Consumer to Prosumer: A Supply Chain Revolution in 3D Printing, International Journal of Physical Distribution & Logistics Management, ISSN 0960-0035, Emerald Publishing Limited, 2018. https://doi.org/10.1108/IJPDLM-03-2018-0139

[6] Nazir, A., Azhar, A., Nazir, U., Liu, Y-F., Qureshi, W.S., Chen, J-E., Alanazi, E., The Rise of 3D Printing Entangled with Smart Computer-Aided Design During COVID-19 Era, Journal of Manufacturing Systems, ISSN 0278-6125, Elsevier Ltd., 2020. https://doi.org/10.1016/j.jmsy.2020.10.009

[7] Troxler, P., van Woensel, C., Socio-Technical Changes Brought About by Three-Dimensional Printing Technology, Proceedings of the Digital Fabrication Conference, ISBN 978-3-8396-0530-9, Germany, November 2013.

Innovative Manufacturing Engineering and Energy - IMANEE2025   Materials Research Forum LLC
Materials Research Proceedings 61 (2026) 17-27   https://doi.org/10.21741/9781644903995-3

[8] Dhir, A., Talwar, S., Islam, N., Alghafes, R., Badghish, S., Different strokes for different folks: Comparative analysis of 3D printing in large, medium and small firms, Technovation, ISSN 0166-4972, Elsevier Ltd., 2023. https://doi.org/10.1016/j.technovation.2023.102792

[9] Ben-Ner, A., Siemsen, E., Decentralization and Localization of Production: The Organizational and Economic Consequences of Additive Manufacturing (3D Printing), California Management Review, ISSN 0008-1256, SAGE Publications, 2017. https://doi.org/10.1177/0008125617695284

[10] Schniederjans, D.G., Adoption of 3D-printing technologies in manufacturing: A survey analysis, International Journal of Production Economics, ISSN 0925-5273, Elsevier B.V., 2017. https://doi.org/10.1016/j.ijpe.2016.11.008

[11] Li, P., Mellor, S., Griffin, J., Waelde, C., Hao, L., Everson, R., Intellectual Property and 3D Printing: A Case Study on 3D Chocolate Printing, Journal of Intellectual Property Law & Practice, ISSN 1747-1532, Oxford University Press, 2014. https://doi.org/10.1093/jiplp/jpt217

[12] Villanueva Orbaiz, M.L., Arce-Urriza, M., The role of active and passive resistance in new technology adoption by final consumers: The case of 3D printing, Technology in Society, ISSN 0160-791X, Elsevier Ltd., 2024. https://doi.org/10.1016/j.techsoc.2024.102500

[13] Mathieson, K., Peacock, E., Chin, W.W., Extending the Technology Acceptance Model: The Influence of Perceived User Resources, Data Base for Advances in Information Systems, ISSN 1532-0936, Association for Computing Machinery (ACM), 2001. https://doi.org/10.1145/506724.506730

[14] Mathieson, K., Peacock, E., Chin, W.W., Extending the Technology Acceptance Model: The Influence of Perceived User Resources, Data Base for Advances in Information Systems, ISSN 1532-0936, Association for Computing Machinery (ACM), 2001. https://doi.org/10.1145/506724.506730

[15] Chen, Z., The Service-oriented manufacturing Mode based on 3D printing: A Case of Personalized Toy, Procedia Engineering, ISSN 1877-7058, Elsevier Ltd., 2017. https://doi.org/10.1016/j.proeng.2017.01.278

[16] Nie, J., Xu, X., Yue, X., Guo, Q., Zhou, Y., Less is more: A strategic analysis of 3D printing with limited capacity, International Journal of Production Economics, ISSN 0925-5273, Elsevier B.V., 2023. https://doi.org/10.1016/j.ijpe.2023.108816

[17] Javaid, M., Haleem, A., Singh, R.P., Suman, R., Rab, S., Role of Additive Manufacturing Applications Towards Environmental Sustainability, Advanced Industrial and Engineering Polymer Research, ISSN 2542-5048, Elsevier Ltd., 2021. https://doi.org/10.1016/j.aiepr.2021.07.005

[18] van Bracht, R., Kleer, R., Piller, F.T., Predicting the Future of Additive Manufacturing: A Delphi Study on Economic and Societal Implications of 3D Printing for 2030, Technological Forecasting & Social Change, ISSN 0040-1625, Elsevier, 2017.

[19] Kim, L., Maijan, P., Yeo, S.F., Developing Customer Service Quality: Influences of Job Stress and Management Process Alignment in the Banking Industry, Sustainable Futures, ISSN 2666-1888, Elsevier Ltd., 2024. https://doi.org/10.1016/j.sftr.2024.100311

[20] Mikula, K., Skrzypczak, D., Izydorczyk, G., Warchoł, J., Moustakas, K., Chojnacka, K., Witek-Krowiak, A., 3D Printing Filament as a Second Life of Waste Plastics-A Review, Environmental Science and Pollution Research, ISSN 0944-1344, Springer, 2021. https://doi.org/10.1007/s11356-020-10657-8

Innovative Manufacturing Engineering and Energy - IMANEE2025
Materials Research Proceedings 61 (2026) 28-36

Materials Research Forum LLC
https://doi.org/10.21741/9781644903995-4

Aspects concerning the microhardness evolution of aluminium ALSI7 welded by MIG and TIG welding technologies

Mihai BOCA[1,a *], Gheorghe NAGÎŢ[1,b], Traian MIHORDEA[2,c],
Tzvetelin GUEORGUIEV[3,d] and Fabian LUPU[4,e]

[1]Gheorghe Asachi Technical University of Iași, Faculty of Machine Manufacturing and Industrial Management, D. Mangeron Bld., No. 59A, Romania

[2]Welding Romanian Association, Branch Iași, Romania

[3]Technical University of Gabrovo, Bulgaria

[4]Gheorghe Asachi Technical University of Iași, Faculty of Mechanics, D. Mangeron Bld., No. 43, Romania

[a]mihai.boca@academic.tuiasi.ro, [b]gheorghe.nagit@academic.tuiasi.ro,
[c]traianmihordea@yahoo.com, [d]tz.georgiev@tugab.bg, [e]fabian-cezar.lupu@academic.tuiasi.ro

Keywords: Heat-Affected Zone, Fusion Welding, Butt Joint Welding, Hardness Evaluation, Hardness Characteristics

Abstract. In all welded assemblies, the joining area must meet the mechanical and functional requirements imposed on the assembly. High importance is placed on the hardness profile across the weld cross-section to determine the mechanical properties of the welded assembly. This study highlights the variation in hardness measured along the cross-section of a welded area in TIG- and MIG-welded AlSi7 joints. The results reveal aspects of grain orientation and suggest that the material may crack outside the weld zone.

Introduction

Welding is a fundamental process in manufacturing and construction, enabling the joining of materials to create complex structures and components. Among the various welding techniques available, Metal Inert Gas (MIG) and Tungsten Inert Gas (TIG) welding are particularly effective for welding aluminium. Aluminium is known for its low density and high strength-to-weight ratio and is widely used in industries such as aerospace, automotive, and construction. However, welding aluminium presents unique challenges due to its high thermal conductivity and the presence of an oxide layer. This article provides a comprehensive comparison of MIG and TIG welding processes for aluminium, focusing on their principles, applications, and performance characteristics.

MIG welding, also known as Gas Metal Arc Welding (GMAW), uses a continuous wire electrode fed through a welding gun and an inert gas, typically argon or a mixture of argon and helium, to shield the weld pool from atmospheric contamination. This process is characterised by high deposition rates, making it suitable for large-scale production and applications requiring high productivity.

Recent advancements in MIG welding technology have focused on improving process control and enhancing weld quality. Pulse-On-Pulse technology, for example, allows for better control of heat input and metal transfer, resulting in improved weld appearance and reduced spatter. Cold Metal Transfer (CMT) is another innovation that reduces heat input, making it ideal for welding thin aluminium sheets and minimising distortion. These advancements have expanded the applicability of MIG welding in various industries, particularly in the automotive and aerospace sectors.

Innovative Manufacturing Engineering and Energy - IMANEE2025   Materials Research Forum LLC
Materials Research Proceedings 61 (2026) 28-36   https://doi.org/10.21741/9781644903995-4

Compared with the MIG process, TIG welding (GTAW) uses a non-consumable tungsten electrode to produce the weld. A separate filler rod is used to add material to the weld pool, and an inert gas, typically argon, is used to shield the weld area from contamination. TIG welding is renowned for its precision and ability to produce high-quality, aesthetically pleasing welds. It is particularly favoured for applications requiring fine, detailed work and where weld appearance is critical.

Innovations in TIG welding have focused on improving process control and enhancing weld quality. Advanced control systems have been developed to manage heat input more effectively, reducing the risk of defects such as porosity and cracking. Techniques such as Activated TIG (A-TIG) welding have been introduced to improve weld penetration and productivity. A-TIG welding involves applying a thin layer of activating flux to the material surface, thereby enhancing arc stability and increasing penetration depth. These advancements have made TIG welding a preferred choice for high-precision applications in industries such as aerospace, medical devices, and electronics.

Numerous studies have been conducted to optimise the MIG welding process for aluminium, focusing on the effects of various welding parameters on weld quality. Research to date has shown that parameters such as welding current, voltage, wire feed speed, and gas flow rate significantly influence the mechanical properties and appearance of the weld. For instance, higher welding currents can increase penetration depth but may also lead to increased spatter and distortion.

Advanced techniques, such as Pulse-On-Pulse MIG welding, have been shown to improve weld quality by providing greater control over heat input and metal transfer. This technique produces smoother weld beads and reduced distortion, making it suitable for applications that require high-quality welds. Additionally, Cold Metal Transfer (CMT) has been explored for welding thin aluminium sheets, as it minimises heat input and reduces the risk of burn-through [1].

Optimisation studies employing methods such as Taguchi have been conducted to identify the optimal combination of welding parameters to achieve desired weld characteristics [2]. These studies have provided valuable insights into the complex interactions between welding parameters and weld quality, thereby enabling the development of more efficient and reliable aluminium welding processes.

Technical literature on TIG welding of aluminium emphasises the importance of precise control over welding parameters to achieve high-quality welds. Parameters such as welding speed, current, electrode type, and gas flow rate play a crucial role in determining the weld bead geometry, mechanical properties, and overall weld quality [3]. Research has shown that slower welding speeds can improve weld penetration and reduce the risk of defects, but may also increase the risk of overheating and distortion. The use of advanced control systems has been explored to more effectively manage heat input, thereby reducing the risk of defects such as porosity and cracking.

Activated TIG (A-TIG) welding has been studied as a technique to enhance weld penetration and productivity. By applying a thin layer of activating flux to the material's surface, A-TIG welding improves arc stability. It increases penetration depth, making it suitable for welding thicker aluminium sections [4]. Optimisation studies using techniques such as Analysis of Variance (ANOVA) and Response Surface Methodology (RSM) have provided insights into achieving high-quality welds with minimal defects, enabling the development of more efficient and reliable TIG welding processes for aluminium [5].

Other studies [6] explore hybrid welding processes that combine GMAW with different techniques, such as laser welding. The study examines the benefits of hybrid processes, including improved weld quality, higher deposition rates, and greater flexibility in welding applications. The researchers conducted experiments to compare the performance of hybrid welding processes with traditional GMAW. The results showed that hybrid welding processes offer significant advantages over conventional GMAW, making them suitable for complex and high-precision applications.

Innovative Manufacturing Engineering and Energy - IMANEE2025   Materials Research Forum LLC
Materials Research Proceedings 61 (2026) 28-36   https://doi.org/10.21741/9781644903995-4

Also, the study [7] investigated the formation of weld defects (specifically porosity) in 2 mm-thick AA7075-T6 plates and analysed how these defects affected the mechanical performance of the joints. The study found extensive formation of large pores and a dendritic structure within the Fusion Zone (FZ). The authors attributed this primarily to inadequate surface cleaning (only mechanical brushing was used), which failed to remove the oxide layer before full welding. Another study [8] examines the impact of helium-argon mixtures on arc stability, penetration, and overall weld quality. The researchers conducted a series of experiments to compare the performance of helium-argon mixtures with traditional shielding gases. The results showed that helium-argon mixtures enhance arc stability and penetration while reducing shielding gas consumption. This research provides valuable insights into cost-effective and efficient alternatives to shielding gas for GMAW. Ultimately, the author Jin [9] focuses on the development of advanced monitoring and control systems for GMAW. The study utilises sensors and machine learning algorithms to provide real-time feedback on the welding process. The researchers demonstrated that these systems can detect and correct defects during welding, thereby improving weld quality and consistency. The methodology involved integrating sensors into the welding equipment and using machine learning models to analyse the data and make real-time adjustments.

Another study by the author Jambor [10] investigates the effects of various gas metal arc welding (GMAW) parameters on the mechanical and microstructural properties of 3 mm-thick S960MC TMCP steel sheets. The research focuses on how varying heat inputs affect the Heat Affected Zone (HAZ), particularly the intercritical HAZ (ICHAZ), which is identified as the weakest region. Despite differences in welding parameters, both welds exhibited significant reductions in tensile, yield, and elongation strengths. The study concludes that conventional GMAW is unsuitable for welding thin TMCP steel sheets due to their susceptibility to overheating and property degradation, suggesting high-energy beam welding as a more suitable alternative. Also, the study performed by Jambor [11] investigates the weldability of S960MC, a thermo-mechanically controlled processed (TMCP) steel with a minimum yield strength of 960 MPa. The authors utilised GMAW on three mm-thick sheets to evaluate how variations in welding parameters (specifically, heat input) influence morphological transformations within the HAZ and the subsequent mechanical performance of the joint. The study concludes that the softening in the Intercritical HAZ fundamentally limits the joint's load-bearing capacity, suggesting that the design allowable must account for this inevitable reduction in local strength. In Cheng [12], a comprehensive review of real-time sensing technologies in GMAW is presented. The authors frame the manufacturing process into three stages: Pre-design, Design (parameter optimisation), and Real-time Control. The study focuses critically on the third stage, addressing the inevitable gap between nominal design conditions and actual manufacturing conditions. The core objective is to minimise welding output deviation through active sensing and feedback control.

In this context, an experimental study of both the mechanical resistance and the microhardness of the welded seam is needed to confirm the mechanical characteristics and the quality of the welded assembly. Thus, the purpose of this study is to develop a perspective on the mechanical characteristics of welded joints in aluminium alloy ALSi7, obtained using both joining welding processes.

Material and experimental methodology
In this study, an ALSi7 aluminium alloy was used. This kind of material has multiple applications. Its mechanical characteristics can be improved by heat treatment, but only when magnesium is present. It is usually used in the automotive industry for wheel fabrication and in the aerospace industry to produce mechanical components with lower mass.

The experiment examines two types of welded samples, with two samples per welding process, focusing on their microstructure and hardness distribution along the transverse direction of the weld. Both samples were obtained by MIG and TIG welding.

Innovative Manufacturing Engineering and Energy - IMANEE2025   Materials Research Forum LLC
Materials Research Proceedings 61 (2026) 28-36   https://doi.org/10.21741/9781644903995-4

The chemical composition of the material, according to EN1706:2020, includes: 6.6-7.5 % Si; 0.15 % Fe; 0.02 % Cu; 0.10 % Mn; 0.30-0.45 % Mg; 0.07 % Zn; 0.10-0.18 % Ti and the aluminium base. The tests were performed on samples collected from an aluminium wheel. The sample preparation was followed: the lateral edges were processed into a V-shaped configuration with an angle of $\alpha = 60°$, a joint opening of r = 2.5 mm, and a joint wall thickness of c = 1.5 ± 0.2 mm. The testing equipment used was a Universal Testing Machine (type WDW-502), manufactured by MCE.

The welding process was carried out in accordance with the work procedure specification established by the welding engineer. Thus, for the TIG process were chosen: the thickness of material was 5mm, as filler material was used an AlSi5 (ER 4043), with a diameter of 2.4 mm, for shielding protection gas was used Argon 4.8 (I1), a preheat temperature between 140-150 °C, the current was alternating current, with an intensity of 80 A, the travel speed was 12-14 cm/min, a butt weld (BW) was chosen as joint type, the grove was type V, without backing, with back gouging and sealing run and the welding position was flat (PA). As a welding source, an EWM Triton 220 AC/DC was used.

For the TIG process, the same AlSi5 alloy was used as the filler material, with a diameter of 1 mm. The current type was Direct Current (DC+), with an intensity of 85 A, and a welding speed of 28-30 cm/min. All other parameters were kept constant. As the welding power source, a DC inverter Type Pico MIG 180 from EWM was used.

After each welding layer was deposited, a visual inspection was conducted, followed by testing with penetrating liquids in accordance with CEN ISO/TR 16060:2015. This step was implemented to assess the weld's susceptibility to potential manufacturing defects over time. No imperfections were identified during intermediate inspections, and the final examination confirmed the absence of any defects.

The evolution of the microstructure after welding was characterised using optical microscopy. Samples were prepared following standard metallographic procedures and etched in Keller's reagent (100 ml; 95 ml H2O, 2.5 ml HNO3, 1.5 ml HCl, 1 ml HF) for 15 seconds at 22 °C, with natural cooling.

Microhardness measurements were utilised to characterise changes in properties across the welds. Microhardness was measured along the midline of a welded nugget, from the base metal through the weld zone to the base metal on the opposite side. For all microhardness measurements, a force of 0.1 kPa (0.98 N) was applied.

Fig. 1 presents a typical balanced welded area, with a coarse-grained region near the fusion line and refined grains outward. The metallic inclusions within the weld metal exhibit optical contrast, indicating their presence. In this case, the inclusions act as stress concentrators, reducing fatigue life and potentially serving as crack initiation sites under cyclic or impact loading.

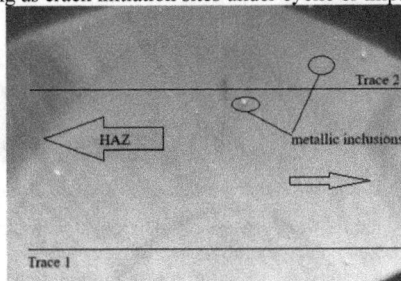

Fig. 1 A perspective concerning the welded seam, with an X-welded profile, MIG Macrostructure.

Also referred to as bead asymmetry, this condition occurs when one side's bead is larger; residual stresses and angular distortion may bias the pass toward the first, larger bead. This can shift peak tensile residual stress into one HAZ, affecting crack growth paths.

Fig. 2 presents a careful examination of the welded area; in this case, the TIG process was free of manufacturing defects and had fewer metal inclusions beyond the demarcation line of the welded area.

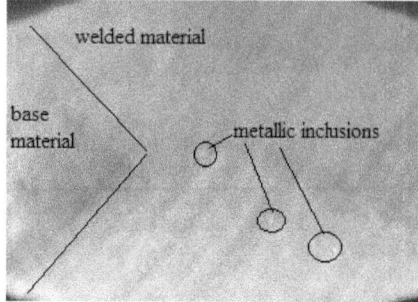

Fig. 2 The TIG microstructure.

The difference between Fig. 1 and Fig. 2 lies in the method used to weld the seam on one side of the joined workpieces.

Fig. 3 presents a typical structure of aluminium, with rapid solidification from the liquid/molten phase, as occurs during welding. Additionally, a fine structure is observed, with small dendrites and equiaxed grains, indicating a high cooling rate. The absence of large grain boundaries suggests that the studied material is a heat-treatable alloy in which phases have precipitated finely. Regarding homogeneity, the structure appears relatively homogeneous, indicating good mixing of the base metal with the filler material.

Fig. 4 reveals a characteristic structure of elongated grains and oriented dendritic cells, indicating preferential directional solidification of the melt pool.

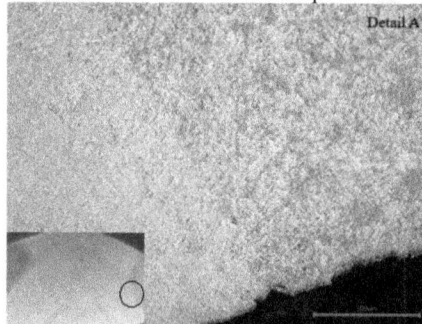

Fig. 3 The MIG microstructure scaled at 500μm.

This fact is explained by the welding process, in which heat is dissipated perpendicular to the fusion line, thereby dictating crystal growth. As we move towards the lower-right corner, the structure becomes less organised and possibly finer, transitioning to an equiaxed structure or the base metal structure, suggesting a transition zone or a change in the cooling rate. Additionally, the

Innovative Manufacturing Engineering and Energy - IMANEE2025   Materials Research Forum LLC
Materials Research Proceedings 61 (2026) 28-36   https://doi.org/10.21741/9781644903995-4

HAZ typically appears as an area in which the strengthening precipitates (the phase that gives the base metal its strength) have dissolved or grown uncontrollably due to welding heat.

Fig. 4 The TIG microstructure scaled at 200μm.

Fig. 5 shows a Vickers hardness test conducted in accordance with ISO 9015-2:2016. The measured value is relatively low, suggesting a soft region—possibly an annealed zone or an austenitic weld metal. The measured lengths of the two diagonals differ slightly, which may reflect minor anisotropies or surface irregularities. The applied load was 200 g, corresponding to 1.96N (HV 0.2), with a dwell time of 5 seconds. The microhardness measurements were performed along two distinct parallel lines on the cross-section: Line 1-First trace, located 1.5 mm below the top surface (face) and Line 2 – Second trace, located 1.5 mm above the root.

Thus, the Vickers microhardness profile was determined by performing two parallel indentation lines across the weld cross-section. The first trace was positioned 1.5 mm from the upper surface, and the second trace was 1.5 mm from the root, with a vertical spacing of 2 mm between the lines. The indentations were spaced 0.5 mm apart along the x-axis.

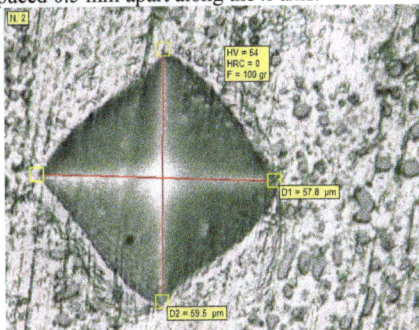

Fig. 5 A detailed aspect of microhardness testing.

Analysing Fig. 6, it can be observed that the microhardness value starts around 40 units, fluctuates mildly, and dips slightly below 40 in the middle region. Initial values may suggest a softer region, typical of unaltered material, which correlates with the measured dip, and may indicate grain coarsening due to thermal exposure, reducing hardness. A slight recovery toward the end could indicate the presence of finer grains or solid-solution strengthening.

Innovative Manufacturing Engineering and Energy - IMANEE2025   Materials Research Forum LLC
Materials Research Proceedings 61 (2026) 28-36   https://doi.org/10.21741/9781644903995-4

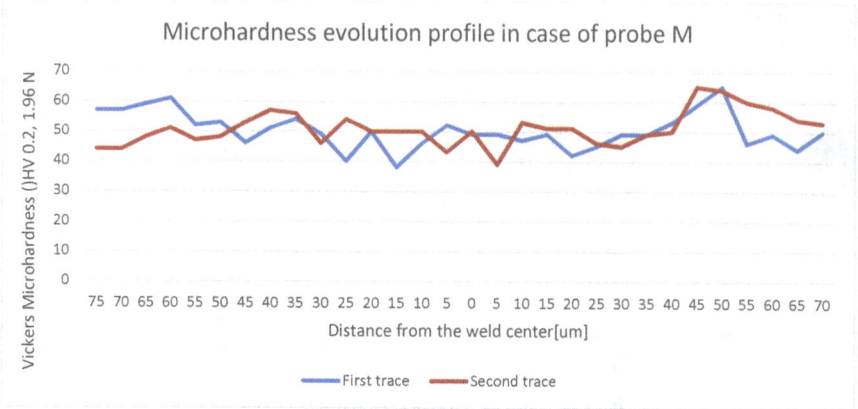

Fig. 6 *Microevolution profile in the case of the MIG sample.*

The second trace, with noticeable peaks, reaches 65 units, which may reflect rapid solidification, leading to fine dendritic grains or alloy segregation that increases hardness. It could be considered that this trace was taken closer to the weld bead; it might show effects of filler metal composition or localised hardening.

In the graph shown in Fig. 7, the blue line indicates a region with uniform microstructure — possibly the base metal or a well-controlled weld zone. At the same time, this aspect suggests minimal thermal damage or grain coarsening, indicating favourable welding parameters and a compatible filler metal. For the red-line path, the profile shows dips below 40 units, suggesting that grain size increases and phase transformation reduces hardness.

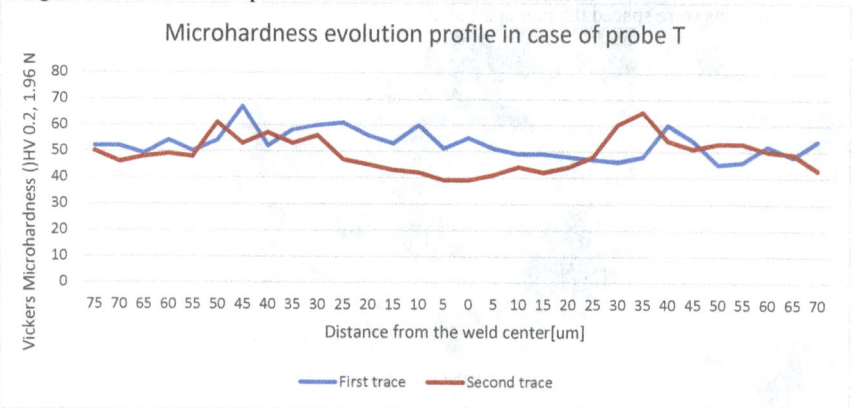

Fig. 7 *Microevolution profile in the case of the TIG sample.*

A possible explanation is that the filler material is more ductile or has a lower alloy content. Additionally, repeated heating and cooling cycles can soften certain regions, particularly in multi-pass welds.

Innovative Manufacturing Engineering and Energy - IMANEE2025
Materials Research Proceedings 61 (2026) 28-36

Materials Research Forum LLC
https://doi.org/10.21741/9781644903995-4

Results and Discussion

The performed tests on tensile strength and bending, concerning the mechanical characteristics of the considered material, allow the following conclusions:

- bending test: fracture/cracking occurred at a location significantly distant from the welded seam;
- thermal influence: the temperature effects in the HAZ for samples did not modify the mechanical resistance properties of the samples;
- micro-cracks: no micro-cracks were observed in the specimens during the bending resistance assessment;
- experimental tests were executed upon the initiation of plastic deformation in each considered sample. In both cases, the deformation was approximately proportional to the applied force.
- the tests were stopped when a decrease in force value was noted, due to the appearance and development of fracture cracks. The curves presented in this study show that the development of deformations under tensile stress has approximately the same shape.

Conclusions

GMAW is a flexible, widely used welding method across diverse industries. Its reliable performance, compatibility with a wide range of materials, and suitability for use in automation systems make it essential in scientific research and experimental manufacturing. The fundamental principles, process settings, and benefits make it a preferred option for numerous welding tasks. Although challenges remain, continuous improvements in GMAW technology are expanding its functionalities and application scope.

The Vickers microhardness values consistently fall within a narrow, uniform range of 40-60 HV, indicating a highly homogeneous hardness distribution across the tested specimens.

This tight interval not only highlights the effectiveness of the processing parameters in reducing microstructural variability but also offers a reliable baseline for comparing wear resistance, mechanical performance, and the effects of subsequent thermal or surface treatments.

Both processes produce virtually indistinguishable hardness profiles, with microhardness values converging into the same narrow band (40–60 HV) across all weld regions. This alignment in hardness indicates that, despite differences in energy input and thermal cycles, both methods drive the microstructure toward an equivalent balance of strength and ductility. In practice, engineers can select either process interchangeably when targeting a specific hardness-driven performance envelope—such as wear resistance, fatigue life, or load-bearing capacity—without sacrificing component uniformity or introducing local soft or hard spots.

Future research and development efforts will likely focus on further improving process efficiency, weld quality, and automation to meet the welding industry's evolving needs.

References

[1] S.-J. Jia, Q.-L. Ma, Y. Hou, B. Li, H.-S. Zhang, Q.-Y. Liu, Changes in microstructure and properties of weld heat-affected zone of high-strength low-alloy steel, J. Iron Steel Res. Int. 31 (2024) 2041–2052. https://doi.org/10.1007/s42243-023-01133-x

[2] C. F. Schenider, C. P. Lisboa, R. Silva, R. T. Lerman, Optimising the parameters of TIG-MIG/MAG hybrid welding on the geometry of bead welding using the Taguchi method, J. Manuf. Mater. Process 1(14) (2017). https://doi:10.3390/jmmp1020014

[3] H. Ji et al., Experimental study on TIG full welding and single pass welding quality of 6061-T6 aluminium alloy sheet, J. of Mat. Res. and Techn. 27 (2023) 5965–5976.

[4] V. Kumar, B. Lucas, D. Howse, G. Melton, S. Raghunnathan, L. Vilarinho, Investigation of an A-TIG Mechanism and the productivity benefits in TIG welding, The 15th Int. Conf. of Joining of Mat. (JOM 15) (2009).

Innovative Manufacturing Engineering and Energy - IMANEE2025   Materials Research Forum LLC
Materials Research Proceedings 61 (2026) 28-36   https://doi.org/10.21741/9781644903995-4

[5] B. Yelamasetti, M. Sandeep, S. S. Narella, V. V. Tiruchanur et al., Optimisation of TIG welding process parameters using Taguchi technique for the joining of dissimilar metals of AA5083 and AA7075, Scientific Reports, 14:23694 2024 https://doi.org/10.1038/s41598-024-74458-6

[6] C.J. Martinez-Gonzalez, A. Barreda, J. Piccini et al., Effects of laser-arc distance on the Process stability and weld quality in hybrid laser-GMAW welding of AISI 304 stainless steel, Metals 9/ 2 (2019).

[7] G. Ipekoğlu and G. Çam, Formation of weld defects in cold metal transfer arc welded 7075-T6 plates and their effect on joint performance, IOP Conf. Series: Mat. Scie. and Eng. 629 012007 (2019), IOP Publishing, https://doi.org 10.1088/1757-899X/629/1/012007

[8] B. J. Kutelu et al, Review of GTAW Welding parameters, J of Mineral Charact. and Eng. 6 (2018) 541-554.

[9] C. Jin and S. Rhee, Real-time weld gap monitoring and quality control algorithm during weaving flux-cored arc welding using deep learning, Metals, 11(7) 11-35 (2021), https://doi.org/10.3390/met11071135

[10] M. Mician, M. Fratrik, D. Kajanek, Influence of welding parameters and filler material on the mechanical properties of HSLA Steel S960MC welded joints, Metals, 11(2) (2021) https://doi.org/10.3390/met11020305

[11] M. Jambor, F. Novy, M. Mician, et al. Gas metal arc welding of thermo-mechanically controlled processed S960MC steel thin sheets with different welding parameters, Communications 20/4 (2018).

[12] Y. Cheng et al., Real-time sensing of gas metal arc welding process – A literature review and analysis, J. of Manuf. Proc. 70 (2021) 452–469.

Innovative Manufacturing Engineering and Energy - IMANEE2025
Materials Research Proceedings 61 (2026) 37-42

Materials Research Forum LLC
https://doi.org/10.21741/9781644903995-5

Influence of printing parameters on overhang structures in micro DLP printing

Martin MICHALAK[1,a], Alexandru SOVER[1,b*], Michael S.J. WALTER[1,c]

[1]Ansbach University of Applied Sciences, Residenzstraße 8, 91522 Ansbach, Germany

[a]martin.michalak@hs-ansbach.de, [b]a.sover@hs-ansbach.de, [c]michael.walter@hs-ansbach.de

Keywords: Micro Printing, Digital Light Processing, Microfabrication, UV Resin

Abstract. Printing of polymer parts via digital light processing (DLP) is an established technology to produce functional complex parts within a short period of time. Within recent years this technology has advanced in even higher geometrical accuracy of the parts since printers and materials were constantly improved. Nowadays, it is possible to print parts with features in micro- (µm) or even nanoscale (nm), opening new fields of application and further improvement to already existing parts. Similar to stereolithography (SLA), DLP applies UV light to harden a photopolymer resin in a layer-by-layer process. This enables printing parts with complex geometries such as cavities, implying the resin can later exit the part. However, DLP has an advantage over other 3D-printing technologies when it comes to creating cavities or the printing of overhang structures in general. Since in DLP the entire layer is hardened at once, less support structure is needed to achieve a uniform, stable layer. In consequence, this opens the design option to print micro cavities without any support structures needed. This paper highlights the process of micro DLP printing of polymer parts, focusing on the influence of the most relevant parameters on the successful print of overhang structures. Therefore, the paper presents an experimental analysis, investigating the effects of relevant printing parameters of an overhang on the dimensional accuracy of the resulting parts. The results show a significant dependence of the bridges (overhangs) on gaps size and curing time, and the deformation of the bridges are in the opposite sides of the parts.

Introduction

Digital light processing is a common method of producing polymer parts based on curing photopolymer resin with UV-light and one of the earliest 3D printing approaches [1]. Parts are built up layer by layer, while submerged in liquid resin, with a projector hardening the resin. After each layer the build plate is moved, and new resin flows in as a new layer to be hardened. The print is based on a stereolithography (STL) file made from a computer-aided design (CAD) being processed in a slicer to fit the printer's parameters [2]. With the development of 3D printers using high resolution light sources and precise motion of less than 1 µm per step [3] on their axis as well as the development of new materials, the precision of the prints improved and made the printing of parts with microstructures possible. This micro printing, often referred as Pµsl [4] opened new fields of application as Ali et al. [5] proved with the printing of polymer micro needles for the medicine field or Carbajo et al. [6] with Micro-Perforated Panel (MPP) sound absorbers for acoustic applications. However, considering the tolerances, functionality and different material properties of micro parts, the whole printing process became more complex as a slight change of parameters will have a higher impact than on a standard macro dimension print. A very important part of printing with resin is the application of proper support structures [7,8] to be able to realize overhang structures or cavities. In this case micro printing can have an advantage over macro printing as the overhang sizes are much smaller and can be realized without support structures or with much less. In the area of macro printing an approach to reduce support structure can be rotating the parts or stacking schemes [9]. This is not always feasible with micro printing as of the

Innovative Manufacturing Engineering and Energy - IMANEE2025
Materials Research Forum LLC
Materials Research Proceedings 61 (2026) 37-42
https://doi.org/10.21741/9781644903995-5

high precision of the parts which can be influenced by rotating due to different precision in x/y and z direction. Marschner et al. [10] already showed the possibility of printing overhangs in micro dimensions using another technology, the two-photon polymerization. For micro printing with DLP another fitting approach would be different parameters influencing the geometrical stability of the layers. To explore the influence of certain parameters on the possibilities of printing parts with overhangs with less support structure or partially without any, following prints and tests were executed.

Methodology

The fabrication of the testing samples was performed on the Giga25vx printer by NanoDimension, a micro DLP printer with an optical resolution of 7 μm and a possible minimum layer height of 1 μm. The material was the P-900 resin from NanoDimension, a precision micro DLP resin. To determine the influence of certain parameters on the printing of overhangs via micro DLP a design of experiments was made, including three different parameters. The three chosen parameters were the UV-Intensity, the exposure time and the size of the overhang. Two of these parameters can be directly changed at the print settings of the printer while the third one is set by the design used for the testing samples. The Design of the testing samples as shown in Fig. 1 were two base pillars on each side with a dimension of 200 μm x 200 μm x 500 μm being connected by an overhang like a bridge.

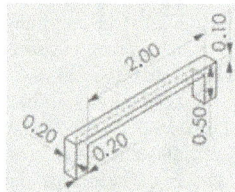

Fig. 1: Design of the testing samples with the set overhang size of 2 mm

For the size of the overhang gap four different lengths were chosen while the width with 200 μm and the thickness of 100 μm stayed the same on all samples. The chosen gape size were 0.5 mm, 1 mm, 1.5 mm and 2 mm as these represent possible gap or overhang sizes for different micro printing scenarios. All samples were printed with a layer height of 10 μm. The sample size for each test was set to five samples which were all printed on the same print (Fig. 2) next to each other to ensure the same quality of the print.

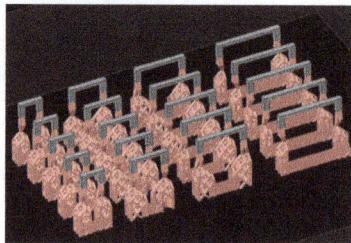

Fig. 2: Samples with support structures in the software Voxeldance Tango

Each print contained all overhang sizes to ensure all of them were printed, which was confirmed by a visual check under the microscope. The support structure was done with the software Voxeldance Tango 4.0 and all the suggested support for the overhang was removed in the software and only the pillars were fitted with support anchors. This was beneficial for the removal of the samples as well, as it made the handling with tweezers easier, being able to grab the support structure and remove it with the sample while later removing the support structure under the microscope. For the UV-Intensity four different levels were chosen and for the exposure time two different levels based on previous printing experience with the same material. With the parameters and the geometry of the samples set, a Design of Experiment was done with the software Minitab to reduce the number of tests needed. As it is a 3 factor with mixed levels design the Taguchi method was used to determine the testing plan shown in Table 1.

Table 1: Design of experiments for the samples

Test number	UV-Intensity in mW	Gap size in mm	UV-Exposure time in ms
V1	35	0.5	360
V2	35	1	360
V3	35	1.5	720
V4	35	2	720
V5	40	0.5	360
V6	40	1	360
V7	40	1.5	720
V8	40	2	720
V9	45	0.5	720
V10	45	1	720
V11	45	1.5	360
V12	45	2	360
V13	50	0.5	720
V14	50	1	720
V15	50	1.5	360
V16	50	2	360

The samples were printed according to the testing plan and afterwards washed with isopropanol before being stored for 24 hours. As results of the tests, the deformation of the bridge was measured via an optical measurement under the microscopy. For this method the two corners of the pillars with the overhang were marked under the Keyence type VHX-7000 microscope with the integrated measurement software and a line was drawn which indicated the middle point. From this middle point a perpendicular of the initial line was put on and the distance between the middle point and the lower side of the overhang was measured as shown in Fig. 3.

Fig. 3: Optical measurement of the sample 2 of the test row 7

To ensure the middle point is the point with the highest deformation, a parallel line of the base line as shown in Fig. 4 was put over the microscopy picture with the software.

Innovative Manufacturing Engineering and Energy - IMANEE2025   Materials Research Forum LLC
Materials Research Proceedings 61 (2026) 37-42   https://doi.org/10.21741/9781644903995-5

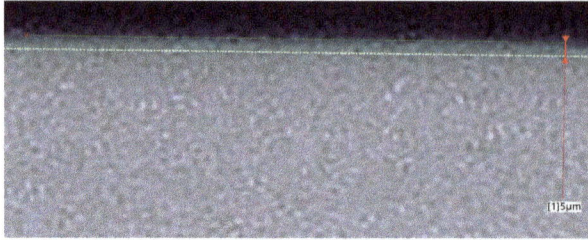

Fig. 4: Distance between base line and parallel line through the highest point of deformation

The measured results of each sample were gathered and later analyzed with Minitab to determine the parameter impacts and the main effects.

Results

The following table 2 shows each measured deformation in μm of each sample as well as the standard deviation of the five samples (S1, S2, …) for every test row (V1, V2, …). All the deformations are in the printing "+z" direction, and therefor only conclude positive numbers.

Table 2: Deformation in μm and the standard deviation

Test	S1	S2	S3	S4	S5	σ
V1	6	6	5	6	6	0.4
V2	19	17	15	14	16	1.72
V3	30	28	28	27	25	1.625
V4	45	50	47	50	49	1.939
V5	7	6	7	7	8	0.632
V6	22	21	18	15	15	2.926
V7	29	27	26	28	27	1.02
V8	43	48	47	42	43	2.417
V9	4	5	4	4	5	0.49
V10	13	11	13	11	12	0.894
V11	39	18	38	36	38	7.96
V12	60	66	61	66	62	2.53
V13	2	2	3	3	2	0.49
V14	6	7	6	7	6	0.49
V15	26	23	25	23	25	1.2
V16	44	41	40	41	41	1.356

Looking at the standard deviations, one stood out, which was the standard deviation of the V11 test row, marked in red, with a value of 7.96 μm. This is more than twice as high as the second highest one marked in yellow with 2.926 μm of the test row V6 indicating that there has been either a measurement error or a sample behaving differently. Sample 2 of the specific test row (marked in orange) showed a very low deformation compared to the rest of the row and as shown in Fig. 4 this was indeed the case with the measurement.

Innovative Manufacturing Engineering and Energy - IMANEE2025   Materials Research Forum LLC
Materials Research Proceedings 61 (2026) 37-42   https://doi.org/10.21741/9781644903995-5

Fig. 5: Optical measurement of the outlier V11, S2

As it is the only sample of all test rows which showed a high deviation from the rest of the row it can be likely that a handling error has occurred. However, due to no other statistical outliers this deviation was not concerning regarding the initial approach. The main effects of the parameters were visualized in the software Minitab and presented in the following Fig. 6.

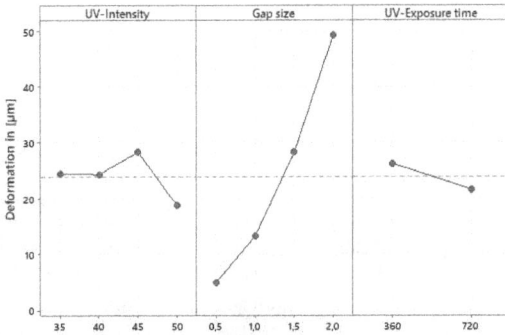

Fig. 6: Main effects of the parameters

While the lower UV-Intensity and exposure time doesn't have a high impact on the deformation, this seems to change with higher intensities and exposure time both leading to lower deformation of the samples. Gaps size on the other hand has the highest impact as the increase in each level of the parameter is rising and overtakes the others in deformation.

Conclusion

The overall results show that it is possible to print certain overhangs without any support structures at all and that a change of certain parameters can influence the deformation in a positive way. The overhangs show a displacement in "+z" direction, which differs to other printing methods e.g. FDM (Fused Filament Fabrication). It seems like the gap size of 2 mm is not the limit and the testing can be extended with bigger gap sizes. However, the results of the deformation of the parts with higher gap size was bigger than expected with the highest deformation of a single sample being 66 µm. This deformation could have occurred due to some post printing curing which is likely as the structure of the overhang was rather weak with only 10 layers and a width of 200 µm. To confirm this suspicion an additional print was done with the same settings as the test row V12, and the samples were measured directly after the cleaning process. The results showed that indeed a deformation while storing took place as the average deformation of the freshly printed samples were 11,8 μm compared to 63 μm of the stored V12 samples. A timelapse under the microscopy then confirmed the slow deformation of the samples over the course of several hours as result of a post print curing. This conclusion is important for further investigating the influences of printing parameters on the possibilities of printing with fewer or none support structure as the geometry of the overhang itself has a great impact. A baseline suggestion for printing overhangs with micro DLP is to try to reduce the size of the overhang and use higher UV-Intensities in a combination with higher exposure time to get the best results possible. A post printing curing process can be necessary depending on the material used and should be determined either by material recommendations or by testing the behavior of the material in a post print timelapse or measurements.

References

[1] S. Swetha, T. J. Sahiti, G. S. Priya, K. Harshitha and A. Anil, Review on digital light processing (DLP) and effect of printing parameters on quality of print, Interactions. 245, 178 (2024). https://doi.org/10.1007/s10751-024-02018-5

[2] P. Fiedor and J. Ortyl A New Approach to Micromachining: High-Precision and Innovative Additive Manufacturing Solutions Based on Photopolymerization Technology, Materials. 13 (2020). https://doi.org/10.3390/ma13132951

[3] C. S. Shin and Y. C. Chang, Fabrication and Compressive Behavior of a Micro-Lattice Composite by High Resolution DLP Stereolithography, Polymers. 13, 785 (2021). https://doi.org/10.3390/polym13050785

[4] Q. Ge, Z. Li, Z. Wang, K. Kowsari, W. Zhang, X. He, J. Zhou and Nicholas X. Fang, Projection micro stereolithography based 3D printing and its applications, Int. J. Extrem. Manuf. 2 (2020). https://doi.org/10.1088/2631-7990/ab8d9a

[5] Z. Ali, E.B. Türeyen, Y. Karpat and M. Çakmakcı, Fabrication of Polymer Micro Needles for Transdermal Drug Delivery System Using DLP Based Projection Stereo-lithography, Procedia CIRP. 42 (2016) 87-90. https://doi.org/10.1016/j.procir.2016.02.194

[6] J. Carbajo, S.-H. Nam and N.X. Fang, Fabrication of Micro-Perforated Panel (MPP) sound absorbers using Digital Light Processing (DLP) 3D printing technology, Applied Acoustics. 216 (2024). https://doi.org/10.1016/j.apacoust.2023.109788

[7] J. Jiang, X. Xu and J. Stringer, Support Structures for Additive Manufacturing: A Review, J. Manuf. Mater. Process. 2 (2018) 64. https://doi.org/10.3390/jmmp2040064

[8] X. Cao, M. Yu, S. Zhang, T. Zhang, Y. Chen, Y. Wang and X. Han, Hierarchical clustering evolutionary tree-support for SLA, J. Manuf. Processes. 125 (2024) 189-201 https://doi.org/10.1016/j.jmapro.2024.07.056

[9] L. Cao, L. Tian, H. Peng, Y. Zhou and L. Lu, Constrained stacking in DLP 3D printing, Computers & Graphics. 95 (2021) 60-69. https://doi.org/10.1016/j.cag.2021.01.003

[10] D. E. Marschner, S. Pagliano, P.-H. Huang and F. Niklaus, A methodology for two-photon polymerization micro 3D printing of objects with long overhanging structures, Additive Manufacturing. 66 (2023). https://doi.org/10.1016/j.addma.2023.103474

Innovative Manufacturing Engineering and Energy - IMANEE2025   Materials Research Forum LLC
Materials Research Proceedings 61 (2026) 43-50   https://doi.org/10.21741/9781644903995-6

Investigations on an efficient computational tool for the simulation of friction stir welding of thermoplastic polymers

Nikolaos E. KARKALOS[1,a] *

[1]Section of Manufacturing Technology, School of Mechanical Engineering, National Technical University of Athens, Heroon Polytechniou 9, 15780 Athens, Greece

[a]nkark@mail.ntua.gr

Keywords: Finite Element Method (FEM), Friction Stir Welding (FSW), Thermoplastic Polymers, Thermal Model

Abstract. Welding is one of the major categories of manufacturing technologies, serving for the joining purposes for a wide variety of workpieces with different shapes and materials. Due to the intense thermo-mechanical phenomena occurring during welding, affecting the workpieces both in the macroscopic and the microscopic level, the choice of the processing conditions is not a trivial procedure in any case. Despite the availability of practical suggestions for the welding of large variety of materials, the increase of the use of some types of materials and the use of novel materials in the industrial applications, the detailed study of the phenomena occurring during welding is necessary. Although a considerable amount of numerical studies on the simulation of friction stir processing already exists, it is necessary to evaluate the validity of low cost and robust models in order to obtain fast but reliable results in the case of process optimization or preliminary studies. In this work, a low cost FE model is implemented and validated against experimental data regarding the prediction of temperature field during the friction stir welding of thermoplastic polymers. The findings of this work showed that although the low cost model contains several simplifications that restrict its accuracy, it was able to predict the temperature field and some heat transfer related quantities reasonably, indicating a potential for further improvement in order to adapt to a wide range of materials and operating conditions.

Introduction

Welding processes are considered a key category of manufacturing techniques, playing an essential role in numerous industrial fields. Their importance stems from the fundamental need to connect separate parts to create complex mechanical systems and structures, which are vital in industries such as automotive, aerospace, construction, shipbuilding, and machinery production. The wide range of welding methods offers flexibility in producing components of various shapes, sizes, and material characteristics, ensuring efficiency in contemporary manufacturing. Consequently, welding remains a critical component in fulfilling the increasing demand for advanced engineering solutions and high-performance products in today's industrial environment.

Especially, Friction Stir Welding (FSW) is a relatively newer solid-state welding method which is employed to join two surfaces with a non-consumable rotational tool without melting but due to mechanical intermixture. This process is considered a serious alterative to more established welding processes as it has various particular capabilities, including the capability of welding materials with low weldability or dissimilar materials in different configurations [1, 2]. Usually aluminum, copper, titanium and magnesium alloys are welded, as well as some steel types and also polymers [3, 4]. Although in general, it is possible to study this process by experimental methods, as different temperature measurement techniques are available and detailed microstructure analysis can help to gain a deep insight into the subsurface alterations of the welded materials, it is not possible to obtain sufficient information about the evolution of the temperature,

Innovative Manufacturing Engineering and Energy - IMANEE2025   Materials Research Forum LLC
Materials Research Proceedings 61 (2026) 43-50   https://doi.org/10.21741/9781644903995-6

stress or strain field within the workpiece as well as the microstructural changes in order to determine possible reasons for defects or other unwanted results.

Throughout the years, models based on heat transfer, fluid dynamics or structural mechanics have been developed for friction stir welding process [5]. Especially, when FEM is used, different types of models have been tested, due to the dynamic nature of the process whereas also mesh-less approaches have been employed in order to avoid distortion due to the high deformation of the elements. Chen and Kovacevic [6] were among the first to use a thermo-mechanical model for the analysis of friction stir welding process. In their work, they took into account the mechanical forces during the process and the thermal phenomena using relevant heat sources and predicted the thermal history and residual stresses of the workpiece, which were compared to the results of experiments. Zhang and Zhang [7] also developed a coupled thermo-mechanical model in order to study the effect of parameters such as the rotational speed in different areas of the workpiece. Aziz et al. [8] included all the FSW steps in their model namely plunge, dwell and travel and studied the effect of different speeds during the different steps of the process, especially on the temperature profile, frictional and dissipation energy. Meyghani et al [9] pointed out the differences between various formulations for the FE model, namely Lagrangian, Eulerian and Arbitrary Eulerian-Lagrangian and in a later study [10] they carried out the respective investigations to evaluate these approaches, finding that ALE method was superior to the others. Myung et al. [11] also compared Lagrangian, Eulerian and ALE models showing that with the ALE models, the sharp gradients of temperature could be predicted as well as the anticipated properties of the welds. Bhattacharjee et al. [12] used a FE model for the FSW of dissimilar materials such as steel and aluminum using the CEL approach, showing that this approach was capable to predict different types of defects under various process conditions as well.

Dialami et al. [13] compared the use of different friction models for the simulation of FSW process, emphasizing not only on the thermal but also on the mechanical effect. Meyghani et al. [14] also studied the case of friction stir welding of curved plates, in which the tool is required to perform a more complex movement and determined the temperature profile. Dialami et al. [15] included special material models to take into account the microstructure evolution during FSW of magnesium alloy for the prediction of metallurgical alterations in the workpiece, such as the grain size and hardness variation. Yoshikawa et al. [16] adopted a particle-based method, namely Moving Particle Semi-implicit (MPS) method for the FSW process, in order to calculate the advective terms more precisely than mesh-based methods as well as deformation. Pan et al. [17] developed a smoothed particle hydrodynamics (SPH) model using non-Newtonian material model for magnesium alloy in order to predict the temperature distribution, grain size, microhardness and evolution of texture.

In the present work, a low cost computational model for the simulation of FSW process focusing on heat transfer phenomena is evaluated for the case of welding of thermoplastic polymers such as High Density Polyethylene (HDPE). After the model is validated based on experimental data, investigations on the effect of process conditions are carried out in order to determine the capabilities and limitations of the model and identify possible opportunities for further enhancement of the model.

Materials and Methods

The model developed in the present work is based on the Finite Element method and is relevant to the heat transfer phenomena which occur during FSW. In this model, the computational domain is composed of the two plates which will be welded and the FSW tool which consists of the shoulder and a threaded pin. As the two plates are from the same material, half of the domain is necessary to be modeled, thus representing only the retreating or advancing side, depending on the direction of the rotation of the tool. In this model only the travel stage is modeled from the three different process stages. Moreover, the tool movement and rotation is not explicitly included in the model

Innovative Manufacturing Engineering and Energy - IMANEE2025   Materials Research Forum LLC
Materials Research Proceedings 61 (2026) 43-50   https://doi.org/10.21741/9781644903995-6

but the tool remains stationary and the velocity field of the deformed material is taken into account through convection velocity, in an Eulerian frame. In order to model the effect of the frictional contact between tool surfaces, especially shoulder and pin, and the workpiece surface, the heat transfer equation, presented in Eq. 1 is solved in the computational domain:

$$\rho C_p \boldsymbol{u} \cdot \nabla T + \nabla \cdot (-k\nabla T) = Q \tag{1}$$

with ρ denoting the workpiece material density, C_p the specific heat, k the thermal conductivity and Q the heat sources. The boundary conditions of the problem include surface heat sources to represent the heat generated at the interface between the shoulder and the workpiece or the pin and the workpiece, as follows:

$$q_{pin}(T) = \frac{\mu}{\sqrt{3(1+\mu^2)}} r_p \omega \sigma_Y(T) \tag{2}$$

$$q_{shoulder}(T) = \frac{\mu F_n}{A_s} \omega r \tag{3}$$

with μ denoting the friction coefficient, r_p the pin radius, ω the rotational speed, σ_Y the yield strength, F_n the normal force, A_s the contact area of the shoulder and workpiece and r the radial distance from the shoulder center. In this model convection, conduction and radiation heat transfer mechanisms are taken into account due to the contact of the workpiece with the surrounding environment and the heat transfer through the workpiece. Convection and radiation is considered at the upper and bottom surface of the plates. The initial temperature of the workpiece is equal to 293 K, except stated otherwise such as a case with preheated workpiece.

The experimental data used for the validation of the proposed model are obtained from the relevant literature, namely the work of Sheikh-Ahmad et al. and Rehman et al. [18, 19]. The workpiece material, geometry and conditions were kept same as in these works. The material used in the simulations is high density polyethylene (HDPE). For the workpiece material, the fundamental thermophysical properties are summarized in Table 1. As it is essential to define the correlation between yield strength and process temperature, the temperature-dependent property is appropriately defined based on experimental data [20]. For the calculations of radiative heat transfer, material emissivity is assumed as 0.3 based on experimental evidence [21]. Friction coefficient (μ) is assumed as 0.2.

Table 1. Thermophysical properties for HDPE.

Density (kg/m³)	Melting temperature (°C)	Yield strength (MPa)	Thermal conductivity (W/mK)	Specific heat (J/gK)
959	131.4	25	0.28	2.25

The tool geometry includes a composite shoulder containing a steel core of 13 mm in diameter and a Teflon collar of 22 mm in diameter around the core as well as a threaded pin with a 6 mm diameter and a length of 5.5 mm. The tool has a tilt angle of 1°. As for the workpiece geometry, the total dimensions of the modeled workpiece are equal to the experimental ones, namely 100 mm length, 6 mm thickness and 37.5 mm in width (half of the original width as only the one plate is modeled).

For the investigation of the effect of process parameters on the FSW process, cases with different rotational speed values, namely 1000, 1250, 1500 rpm and tool feed rate, namely 20, 30, 40 mm/min were considered, according to the experimental work.

Innovative Manufacturing Engineering and Energy - IMANEE2025
Materials Research Proceedings 61 (2026) 43-50

Materials Research Forum LLC
https://doi.org/10.21741/9781644903995-6

As it is obvious from Eq. 1, the model is steady-state. The size of the mesh is 49726 elements, with the vast majority of them being tetrahedral and a small part being prisms. For the simulations, COMSOL® Multiphysics version 6.2 was used.

Results and Discussion

Investigation on the effect of process parameters. The first step of the evaluation of the model consisted of the comparison with available experimental results. The selected case from an existing study [19] corresponds to the following conditions: rotational speed of 1500 rpm, feed rate of 30 mm/min and initial temperature of 60°C. In that work, the temperature along the weld line at the bottom of the workpieces was measured and the average temperature determined in three different locations along the weld line was around 117°C, whereas the predicted temperature was 117.2°C, indicating a sufficient level of accuracy. Thus, after the validation, further investigations were able to be carried out with the simulation model and the corresponding results were analyzed.

In the case of temperature values, evaluated near the weld line at the bottom surface, as can be seen in Fig. 1, the simulation results are in accordance with most of the relevant studies. The increase of rotational speed leads to a barely noticeable increase of less than one degree Celsius as the frictional energy increases. Although this is a reasonable trend, this result also indicates that the rotational speed is not a very significant parameter for temperature, compared to the feed rate. This observation can be attributed to the fact that the change of rotational speed from 1000 to 1500 rpm is not high but also to one of the main deficiencies of the selected model which does not take the deformed material flow explicitly into account, especially the mixing due to the rotational speed of the tool, underestimating the considerable importance of this parameter. On the other hand, the increase of the feed rate causes a clear decrease of temperature, as can be seen in Fig. 1, as the higher feed rate allows less time for the interaction between tool and workpiece, with lower frictional energy developed and consequently lower heat and lower temperatures in the workpiece.

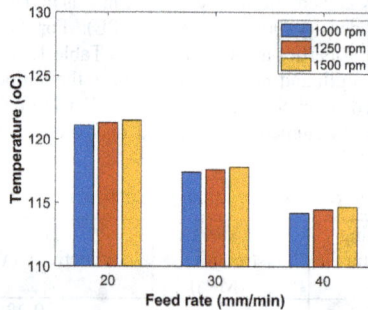

Fig. 1 Variation of predicted temperature in respect to the process conditions.

In order to observe the temperature field, an indicative image of the temperature field is plotted in Fig. 2, for the case with rotational speed of 1000 rpm and feed rate of 20 mm/min.

Innovative Manufacturing Engineering and Energy - IMANEE2025   Materials Research Forum LLC
Materials Research Proceedings 61 (2026) 43-50   https://doi.org/10.21741/9781644903995-6

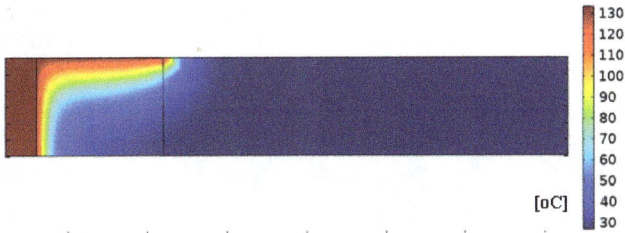

Fig. 2 Temperature field in a cross-section along the workpiece width.

In order to further gain insight into the thermal phenomena near the weld line, two additional quantities are studied, namely the temperature gradient and the extent of the "affected" zone. From the temperature profiles, it was observed that a small region next to the pin exhibited high temperatures, whereas the temperature decreased rapidly away from that area. Thus, the temperature gradient was calculated in a line along the workpiece width at the bottom surface, extending from the tool surface. It was confirmed that the highest value of the temperature gradient was obtained in the region between the pin and the unaffected material zone and the maximum temperature gradient values were calculated, as can be seen in Fig. 3. From this Figure, it can be seen that the values of the temperature gradient vary according to the different rotational speed and feed rate values with different trends than the temperature. More specifically, the temperature gradient increases only slightly when the rotational speed increases but increases more clearly when feed rate increases. Thus, these conditions can affect not only the temperature field but also the sharpness of the gradient, with more abrupt changes in the temperature occurring at high feed rates, when the insufficient amount of heat produced leads to low temperatures close to the weld line, and high rotational speeds.

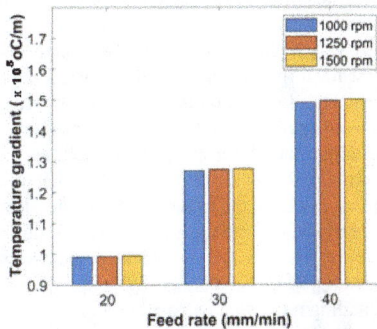

Fig. 3 Variation of predicted temperature gradient in respect to the process conditions.

Finally, although the microstructural alterations and thus the thermo-mechanical affected zone (TMAZ) and heat affected zone (HAZ) cannot be calculated by the present model, the size of the area between the weld line and the unaffected material zone can be estimated by the temperature profile. In Fig. 4, it can be seen that the extend of the "affected" zone, corresponding to the actual TMAZ and HAZ zones, can be varied according to the rotational speed and feed rate values. In specific, the rotational speed leads to a slight widening of the "affected zone" whereas the increase

Innovative Manufacturing Engineering and Energy - IMANEE2025
Materials Research Proceedings 61 (2026) 43-50

Materials Research Forum LLC
https://doi.org/10.21741/9781644903995-6

of the feed rate reduces the width of the "affected" zone considerably as the lower amount of heat produced due to frictional energy can affect a rather small zone near the weld line.

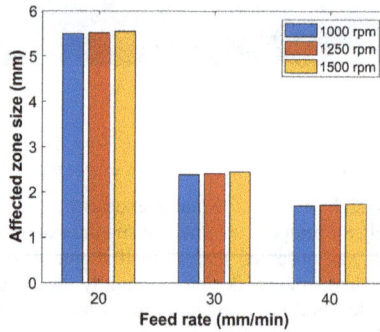

Fig. 4 Variation of the size of the predicted affected zone in respect to the process conditions.

Evaluation of the limitations of the present model. The analysis of the results obtained by the proposed FSW model allowed for the identification of the strengths, limitations and potential improvements for the model.

More specifically, it is expected that this model can be mainly used for preliminary study or optimization purposes, before a more detailed one can be used or experimental results are available. This model includes the real geometry of the workpiece with the exact dimensions without the need for simplifications, and material properties can be directly used, even temperature-dependent ones. A considerable advantage of the model is the reduced computational time and cost, as feed rate which is rather low in the FSW experiments, is not explicitly modeled and also there is no need for remeshing or moving mesh techniques and any problem related to the mesh element distortion is avoided as the tool remains at the same position. Moreover, as most of main heat transfer mechanisms are included the extent of affected zones and temperature gradients can be approximated, at least for a first insight into the process, especially when new workpiece materials are tested. Furthermore, a fairly good representation of temperature field consistent with experimental findings can be obtained and the model can approximate the developed temperatures with sufficient accuracy mainly based on properties such as the melting temperature and temperature-dependent yield strength. On the other hand, the lack of the representation of material deformation and the inability to calculate the mixing phenomenon lead to reduced accuracy for the prediction of the significance of some parameters and does not allow for studying the stress field evolution or the identification of possible defects which can be determined by other methods.

However, this model can be possibly expanded by coupling it with a fluid flow model or a variable friction model and adding more sophisticated models for the prediction of microstructural alterations or hardness, which after appropriate validation will improve the practical use of the model.

Conclusions

In the present work, commercial software was used to implement a low cost FE model for FSW process, focusing mainly on the heat transfer phenomena. After the model was validated, an investigation on the effect of tool rotational speed and feed rate was carried out and various quantities related to the temperature field were analyzed, whereas the capabilities and weaknesses of the model were identified and discussed. From this work several conclusions were drawn:

Materials Research Forum LLC
https://doi.org/10.21741/9781644903995-6

Temperature field is affected by both the rotational speed and feed rate of the tool, to a different extent. The increase of rotational speed and decrease of feed rate lead to an increase of temperature of the workpiece, due to higher frictional work. Although the model underestimates the importance of rotational speed, however it predicts the relevant trends accurately and can provide a reliable preliminary estimation of the consequences of the selected process parameters for the weld.

The analysis of the temperature gradient values indicated a strong correlation with feed rate, with sharper gradients occurring when feed rate is higher and also a noticeable difference in the extent of the affected areas when feed rate varies, as larger parts of the workpiece have elevated temperatures at lower feed rates.

This study allowed for the identification of strengths and discrepancies of the proposed method, which enabled a preliminary study of the heat transfer phenomena during FSW of thermoplastic polymers and the effect of different process parameters but cannot offer a more detailed view on the material mixing and deformation phenomena during FSW process.

References

[1] M.M.Z. Ahmed, M.M.S. Seleman, D. Fydrych, G. Cam, Friction stir welding of aluminum in the aerospace industry: the current progress and state-of-the-art review, Materials 16(8) (2023) 2971. https://doi.org/10.3390/ma16082971

[2] D.G. Mohan, C. Wu, A review on friction stir welding of steels, Chin. J. Mech. Eng. 34 (2021) 137. https://doi.org/10.1186/s10033-021-00655-3

[3] M.M. El-Sayed, A.Y. Shash, M. Abd-Rabou, M.G. ElSherbiny, Welding and processing of metallic materials by using friction stir technique: A review, J. Adv. Join. Process. 3 (2021) 100059. https://doi.org/10.1016/j.jajp.2021.100059

[4] R. Rudrapati, Effects of welding process conditions on friction stir welding of polymer composites: A review, Compos. C: Open Access 8 (2022) 100269. https://doi.org/10.1016/j.jcomc.2022.100269

[5] X. He, F. Gu, A. Ball, A review of numerical analysis of friction stir welding, Progr. Mater. Sci. 65 (2014) 1-66. https://doi.org/10.1016/j.pmatsci.2014.03.003

[6] C.M. Chen, R. Kovacevic, Finite element modeling of friction stir welding - thermal and thermomechanical analysis, Int. J. Mach. Tools Manuf. 43 (2003) 1319-1326. https://doi.org/10.1016/S0890-6955(03)00158-5

[7] Z. Zhang, H.W. Zhang, A fully coupled thermo-mechanical model of friction stir welding, Int. J. Adv. Manuf. Technol. 37 (2008) 279-293. https://doi.org/10.1007/s00170-007-0971-6

[8] S.B. Aziz, M.W. Dewan, D.J. Huggett, M.A. Wahab, A.M. Okeil, T.W. Liao, A fully coupled thermomechanical model of friction stir welding (FSW) and numerical studies on process parameters of lightweight aluminum alloy joints, Acta Metall. Sin. (Engl. Lett.) 31 (2018) 1-18. https://doi.org/10.1007/s40195-017-0658-4

[9] B. Meyghani, M.B. Awang, S.S. Emamian, M.K.B.M. Nor, S.R. Pedapati, A comparison of different finite element methods in the thermal analysis of Friction Stir Welding (FSW), Metals 7(10) (2017), 450. https://doi.org/10.3390/met7100450

[10] B. Meyghani, M.B. Awang, M. Momeni, M. Rynkovskaya, Development of a finite element model for thermal analysis of friction stir welding (FSW), IOP Conf. Ser. Mater. Sci. Eng. 495 (2019) 012101. https://doi.org/10.1088/1757-899X/495/1/012101

[11] D. Myung, W. Noh, J.H. Kim, J. Kong, S.T. Hong, M.G. Lee, Met. Mater. Int. 27 (2021) 650-666. https://doi.org/10.1007/s12540-020-00901-8

[12] R. Bhattacharjee, S. Datta, A. Hammad, P. Biswas, Prediction of various defects and material flow behavior during dissimilar FSW of DH36 shipbuilding steel and marine grade AA5083 using FE-based CEL approach, Model. Simul. Mater. Sci. Eng. 31(3) (2023) 035004. https://doi.org/10.1088/1361-651X/acbe5a

[13] N. Dialami, M. Chiumenti, M. Cervera, A. Segatori, W. Osikowicz, Enhanced friction model for Friction Stir Welding (FSW) analysis: simulation and experimental validation, Int. J. Mech. Sci. 133 (2017) 555-567. https://doi.org/10.1016/j.ijmecsci.2017.09.022

[14] B. Meyghani, M. Awang, C.S. Wu, Finite element modeling of friction stir welding (FSW) on a complex curved plate, J. Adv. Join. Process. 1 (2020) 100007. https://doi.org/10.1016/j.jajp.2020.100007

[15] N. Dialami, M. Cervera, M. Chiumenti, Numerical modeling of microstructure evolution in friction stir welding (FSW), Metals 8(3) (2018) 183. https://doi.org/10.3390/met8030183

[16] G. Yoshikawa, F. Miyasaka, Y. Hirata, Y. Katayama, T. Fuse, Development of numerical simulation model for FSW employing particle method, Sci. Technol. Weld. Join. 17(4) (2012) 255-263. https://doi.org/10.1179/1362171811Y.0000000099

[17] W. Pan, D. Li, A.M. Tartakovsky, S. Ahzi, M. Khraisheh, M. Khaleel, A new smoothed particle hydrodynamics non-Newtonian model for friction stir welding: Process modeling and simulation of microstructure evolution in a magnesium alloy, Int. J. Plast. 48 (2013) 189-204. https://doi.org/10.1016/j.ijplas.2013.02.013

[18] J.Y. Sheikh-Ahmad, S. Deveci, F. Almaskari, R.U. Rehman, Effect of process temperatures on material flow and weld quality in the friction stir welding of high desnity polyethylene, J. Mater. Res. Technol. 18 (2022) 1692-1703. https://doi.org/10.1016/j.jmrt.2022.03.082

[19] R.U. Rehman, J. Sheikh-Ahmad, S. Deveci, Effect of preheating on joint quality in the friction stir welding of bimodal high density polyethylene, Int. J. Adv. Manuf. Technol. 117 (2021) 455-468. https://doi.org/10.1007/s00170-021-07740-w

[20] N. Merah, Z. Khan, A. Bazoune, F. Saghir, Temperature and loading frequency effects on fatigue crack growth in HDPE pipe material, Arab. J. Sci. Eng. 31(2) (2006) 19-30.

[21] T. Okada, R. Ishige, S. Ando, Analysis of thermal radiation properties of polyimide and polymeric materials based on ATR-IR spectroscopy, J. Photopolym. Sci. Technol. 29(2) (2016) 251-254. https://doi.org/10.2494/photopolymer.29.251

Innovative Manufacturing Engineering and Energy - IMANEE2025   Materials Research Forum LLC
Materials Research Proceedings 61 (2026) 51-57   https://doi.org/10.21741/9781644903995-7

Quality assessment considerations in project management of new product development - A case study

Mihaela NICOLAU[1,a *], Roxana Gabriela HOBJÂLĂ[1,a] and Oana DODUN[1,a]

[1]Gheorghe Asachi" Technical University of Iași, Department of Machine Manufacturing Technology, Blvd. D. Mangeron, 39 A, 700050 Iași România

[a]mihaela.nicolau@student.tuiasi.ro, [b]roxana-gabriela.hobjila@student.tuiasi.ro, [c]oana.dodun-des-perrieres@academic.tuiasi.ro

Keywords: Quality, Decision-Making, Ishikawa Diagram, Project Management, Case Study

Abstract. In the development of a new product, plans are put in place to ensure the successful completion of the endeavour. As the elements involved in developing the end result can be numerous and sometimes unknown to some participants, poor communication among team members is almost certain to occur. This will most likely lead to low-quality work they perform together. Many methods have been developed to identify causes of poor results. One of them is the Ishikawa diagram. The paper proposes a quick method to be used in numerous situations that arise in the project management of a NPD. The method relies on Axiomatic Design Theory, which easily breaks down items to a very finite level of information. A case study will be utilised to exemplify the benefits of the method. The results confirm the usefulness of the proposed method for identifying high-risk elements that could arise in the project management of a NPD.

Introduction

Companies frequently develop new products and services in order to remain competitive and to respond effectively to evolving customer needs. Such innovation-driven activities are typically structured and managed as projects, as noted by Greasley [1], because they involve a defined objective, a temporary timeframe, and the coordinated allocation of resources. According to the Project Management Institute (PMI) [2], projects are managed through the systematic integration of multiple knowledge areas and a series of interrelated processes that ensure the successful achievement of the intended outcome—for example, the development of a new product.

Project management practices are organized within a well-established project life cycle, which provides a structured path from conception to completion. The life cycle generally begins with the initiation phase, where the project's purpose, feasibility, and high-level requirements are defined. This is followed by the planning phase, in which the project scope is refined, objectives are detailed, and strategies for schedule, cost, quality, and resources are developed. The execution and monitoring/control phase encompasses the coordination of teams, allocation of work, tracking of progress, and management of potential deviations to ensure alignment with the baseline plan. Finally, the life cycle concludes with the closing phase, where project deliverables are finalized, evaluated, formally accepted, and lessons learned are documented for organizational improvement [3].

To support these phases, PMI identifies ten core knowledge areas that collectively capture the managerial and technical dimensions of a project: Integration, Scope, Schedule, Cost, Quality, Resources, Communications, Risk, Procurement, and Stakeholder Management. Each knowledge area contributes a specific set of tools, processes, and techniques that help project managers ensure coherence, manage complexity, and align the project's deliverables with strategic business goals.

If project management is focused on organizing activities [4] of various departments participating in a project, new product development (NPD) incorporates a series of steps that begin

Innovative Manufacturing Engineering and Energy - IMANEE2025   Materials Research Forum LLC
Materials Research Proceedings 61 (2026) 51-57   https://doi.org/10.21741/9781644903995-7

with ideation [discovery and screening], followed by conceptualization [design, prototyping, and manufacturing], and finishing with road-mapping, and launch [5].

Quality improvement has been the focus of many studies and led to the development of many methodologies such as Quality Function Deployment (QFD), Total Quality Management (TQM), Project Management, and Axiomatic Design (AD) [6] just to name a few.

It will be highly beneficial for a project team to quickly assess the root cause of problems ahead of time while building the risk management plans before they begin developing the product.

Quick research through ScienceDirect database of studies focusing on AD and risk assessment found approximately 2500 articles analyzing this topic with applications from supply chain to preventive medical procedures. But no study could be found connecting AD to any of the problem analysis techniques.

Quality Assessment Techniques

Quality considerations in NPD refer to product quality. PMI focuses on quality management by planning, assuring, and controlling project quality. According to PMI, quality planning uses techniques such as cost benefit analysis, benchmarking, design of experiments, cost of quality, etc. For quality assurance, the techniques are planning tools, audits, process analysis, and control tools [2]. For quality control [2] mentions cause and effect diagram (Fishbone/Ishikawa diagram), control charts, flowcharting, histogram, Pareto charts, run chart, Scatter diagram, statistical sampling, inspection, and defect repair review. In addition to this, ISO/IEC 31010 standard mentions several risk assessment techniques such as Root Cause Analysis (RCA), Failure mode and effects analysis (FMEA), Fault tree analysis, multi-criteria decision analysis (MCDA), etc. [7].

Each of these techniques has been developed to resolve certain problems. Pareto analysis, also known as 80/20, focuses on identifying the most significant issue. It is useful in prioritizing multiple problems. Scatter Diagrams find the correlation between two variables, and it is useful in visualizing data points. Fault tree analysis is another method that breaks down a problem into non-classified categories. It is useful in complex problems [8]. RCA connects the problem to the root causes using factual data. It is used systematically to monitor business processes [9]. FMEA analyzes the functional, design, and process failures that lead to a component failure. It estimates the probability, severity and detection of the cause from appearing. It identifies the major faults in a system [10]. MCDA evaluates multiple conflicting options to make better selections. It can be applied to a wide range of situations for which many techniques have been developed [11].

The Fishbone Diagram (FD), also known as the Ishikawa Diagram or Cause-and-Effect Diagram, is a structured visual tool used to systematically identify, explore, and display the possible causes of a specific problem or outcome. In its classical form, when applied to product failures, the FD decomposes potential sources of defects into well-established categories such as Processes, People, Environment, Materials, Measurements, and Tools. For a failed physical product, these branches capture all technical and operational contributors, while for a failed component, the diagram may further expand into

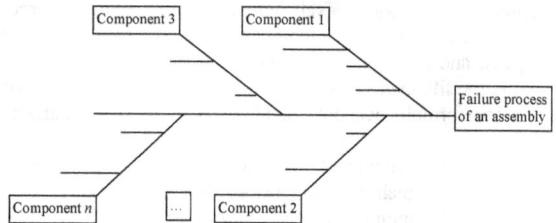

Figure 1 - Assembly failure cause decomposition.

all subassemblies and subcomponents that could influence the failure mechanism (Fig. 1).

However, when the Fishbone Diagram is used within the project management domain, the causal categories must be aligned with the structure of project processes rather than manufacturing or design processes. Therefore, the root causes are typically organized according to the ten knowledge areas defined in project management standards. These include Integration, Scope, Time, Cost, Quality, Human Resources, Communications, Risk, Procurement, and Stakeholders. Each category provides a lens through which deviations, delays, or failures in project performance can be analyzed, enabling a comprehensive evaluation of managerial, organizational, and procedural factors (Fig. 2).

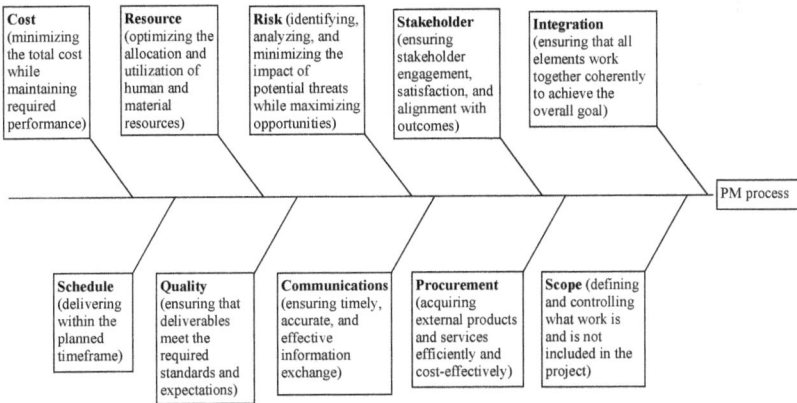

Figure 2 - Project Management process decomposition as an Ishikawa Diagram.

By adapting the causal branches to the specific context, whether a physical product failure or a project management issue, the Fishbone Diagram becomes a flexible and powerful tool for root-cause analysis, supporting decision-making and continuous improvement across technical and managerial environments.

The analysis starts from the most relevant, the primary, cause and continues to the secondary cause, and so on. It is a qualitative analysis tool [12], easy to use, and it can be applied to complex problems.

Axiomatic Design

AD is a systemic approach to decomposing information into domains, mapping them into design matrices and improving the results through axioms [13,14,15,16].

The domains are identified as customer, functional, physical, and process domains and are associated to customer needs (CN) information, product functional requirements (FR) which are derived from CN, product physical design parameters (DP) which are derived from FR, and process variables (PV) which describe the processes achieving DPs [17].

This information is then filtered through two axioms. The first axiom, known as Independence Axiom, requires FRs to be independent, which is means that for each FR there is only one DP, and for each DP there is only one PV. The second axiom, known as Information Axiom, requires the information content to be minimal so that it cannot be further decomposed in smaller units.

The process begins by collecting the CNs. They are translated into FRs. For each FR, a unique DP needs to be identified. For each DP there can be only one process to manufacture PV and test that DP. This is continued until no further decomposition can be performed. This is known as the mapping of domains, and it is represented by a matrix.

The mapping of FRs to DPs results in a product design matrix [A] which can be expressed by Eq. 1:

$$\{FR\} = [A]\,\{DP\} \tag{1}$$

The mapping of DPs to PVs results in a process matrix [B] as expressed by Eq. 2:

$$\{DP\} = [B]\,\{PV\} \tag{2}$$

where:

{FR} is the vector of functional requirements;
{DP} is the vector of the design parameters;
{PV} is the vector of the process variables;
[A] is the design matrix;
[B] is the process matrix.

An ideal design of n functions/components will be represented as shown in Table 1. This is called an uncoupled design, either wise it will be a coupled design, when there are less DPs than FRs, or redundant design, when there are more DPs than FRs [17].

Table 1. A product design matrix

Functional Requirements	Design Parameters		
	DP1	DP2	DPn
FR1	X		
FR2		X	
FRn			X

In the area of NPD, all product physical design parameters are the components subjected to possible failures.

Case Study

In this article, the authors focus on the design of a 3D Printing unit that utilizes readily available components.

The main functions of the unit will be to create a three-dimensional object from a digital model by layering material in a controllable environment.

The main objective of the system will be to add material according to a preexisting design. They form the first level functional requirements and design parameters:

FR1 = add material   DP1 = 3D printing assembly
FR2 = control data   DP2 = controller board module

The material addition function can be further decomposed into providing printing material, providing printing surface, providing motion control, providing power supply, and providing enclosure:

FR1.1 = provide printing material   DP1.1 = nozzle driving system
FR1.2 = provide printing surface   DP1.2 = print bed assembly
FR1.3 = provide motion control   DP1.3 = motion systems
FR1.4 = provide power supply   DP1.4 = power supply
FR1.5 = provide enclosure   DP1.5 = frame

The mechanical components are to be controlled by providing data transfer, human machine communication and deploying a design:

FR2.1 = provide data transfer   DP2.1 = motherboard module
FR2.2 = provide communication control   DP2.2 = user interface module
FR2.3 = deploy the design   DP2.3 = memory storage unit

If these modules are not readily available, there is a need to decompose the design into more refined details.

In providing printing material, several functions need to be performed by certain modules, such as supplying and transporting the cold filament towards the hot nozzle/end.

FR1.1.1 = provide cold filament   DP1.1.1 = filament feeder system
FR1.1.2 = provide heated material   DP1.1.2 = hot end assembly

These modules are to be evaluated for either acquisition or manufacturing for which further decomposition is necessary.

Innovative Manufacturing Engineering and Energy - IMANEE2025   Materials Research Forum LLC
Materials Research Proceedings 61 (2026) 51-57   https://doi.org/10.21741/9781644903995-7

In order to provide a printing surface, additional functions are to be performed such as holding a part on a heated horizontal surface.

FR1.2.1 = provide for rigorous platform   DP1.2.1 = print bed
FR1.2.2 = provide heating   DP1.2.2 = heating system
FR1.2.3 = provide for part adhesion   DP1.2.3 = adhesive sheet
FR1.2.4 = provide for self-leveling   DP1.2.4 = self-leveling system

In regard to providing motion control in either the horizontal or vertical way, further decomposition is necessary such as propelling the hot end in all directions and sensing its position.

FR1.3.1 = propel the hot end   DP1.3.1 = stepper driver module
FR1.3.2 = propel hot end in x and y direction   DP1.3.2 = belt
FR1.3.3 = propel hot end in z direction   DP1.3.3 = leadscrew
FR1.3.4 = detect hot end in all directions   DP1.3.4 = end stop

In general, if none of these components are available for acquisition, then further decomposition is necessary. Assuming that all of these components are at the lowest level of decomposition, we check to see if they satisfy the requirements of axioms 1 and 2 by which the FRs must be independent and sufficient.

From the lowest level moving upwards we build matrices at each level in order to determine if the design forms an uncoupled design matrix, as shown in Table 2.

The X marks the unique correspondence of an FR to a DP. As all functions can be described by unique physical modules, the design is considered complete.

Once the design is considered complete,

Table 2. The 3rd level product design matrix

Functional Requirements 3rd level	Design Parameters of 3rd level			
	DP1.2.1	DP1.2.2	DP1.2.3	DP1.2.4
FR1.2.1	X			
FR1.2.2		X		
FR1.2.3			X	
FR1.2.3				X

the DPs become subjected to potential failure cause analysis to better understand possible risks.

The first step is determining the most influential factor for that failure. In this case the print bed is not working properly, and each subcomponent is considered.

The analysis will look at the influence of processes, people, environment, materials, measurements, and tools as shown in Table 3, but in other applications there might be different factors.

Table 3. DP and Failure Modes

Failure Modes	DP1.2.1	DP1.2.2	DP1.2.3	DP1.2.4
Process	Make	Make	Buy	Make
Environment	InHouse	Supplier	Online	InHouse
Materials	Made	Made	As Is	Made
Machine	As Is	As Is	As Is	As Is
Measurements	As Is	As Is	As Is	As Is
Operator	Tech	Tech	Buyer	Tech

Table 3 depicts how the print bed subassembly can be analyzed for potential failure modes.

Potential sources are identified as any standardized making or buying methods not being followed, any environmental condition variation whether it is inhouse or at a supplier or online, any resource mishandling or of low quality whether that is made inhouse or purchased, any facility or equipment usage, maintenance, or assembly, any quality control standards and techniques, any improper behavior of any person participating tin the development of that component or assembly. These findings can be further refined until no smaller reason can be found to be causing problems.

When all possible sources of failure are identified, an investigation into how they occur is necessary in order to define and prioritize solutions for the current situation and a new approach in the future.

Discussion

Through AD a detailed decomposition of a system results in the identification of product components and subassemblies. The Fishbone Diagram is used to decompose possible failure modes for a problematic component into categories.

This analysis can be made before production begins in order to eliminate endangering the success of the project. Therefore, all components and subassemblies identified in this study could be analyzed in a similar way.

If the results of the analysis lead to the conclusion a redesign is needed, this process becomes a very cost-effective approach.

The combination of these two methods offers clarity and expediency in eliminating possible problems that could have been otherwise dangerous and costly to fix. The method is easily applicable in other areas in which either of the methods would be considered. A larger spreadsheet can develop in order to incorporate as many details as possible as it seems necessary. This allows all members participating in the project to exchange quality project data among themselves.

Conclusion

In new product development there is always an interest in improving the quality of products and processes. Both methods, Axiomatic Design and Fishbone Diagram, are well fit for better defining a design.

Axiomatic Design decomposes customer needs into functional requirements of a product for which physical sub-assemblies, known as design parameters, are identified. It is considered a good design when for each function there is only one physical component. This forms the product structure.

The Fishbone Diagram looks at the product structure from the perspective of the most relevant categories of problems applicable to that situation. This is an effective and structured tool that facilitates the understanding of different factors.

In the case of a 3D printer's bed issues, the categories utilized are processes, people, environment, materials, measurements, and tools. The results show that the approach is structured, fast, scalable and transferable to other industries generating better designs, saving time and money for a successful NPD project management.

References

[1] A. Greasley, Operations Management, SAGE Publications Ltd., London, 2007. https://doi.org/10.4135/9781446213025

[2] Project Management Institute, A Guide to the Project Management Body of Knowledge (PMBOK® Guide), sixth ed., Project Management Institute, Newtown Square, Pennsylvania, 2017.

[3] M. Pessoa, L.G. Trabasso, The Lean Product Design and Development Journey: A Practical View, Springer International Publishing, Cham, 2017, pp. 66–67. https://doi.org/10.1007/978-3-319-46792-4

[4] Ş. Demirel, A. Camci, Project management vs product management: a view on the journey from project to product in software development, in: Proceedings of the International Conference on Industrial Engineering and Operations Management, Istanbul, Turkey, 2022.

[5] B. Bigliardi, E. Bottani, M. Rinaldi, The new product development process in the mechanical industry: evidences from Italian case studies, Int. J. Eng. Sci. Technol. 5(2) (2013) 1–23. https://doi.org/10.4314/ijest.v5i2.1S

[6] Y.H. Wang, C.H. Lee, A.J.C. Trappey, Service design blueprint approach incorporating TRIZ and service QFD for a meal ordering system: a case study, Comput. Ind. Eng. 107 (2017) 388–400. https://doi.org/10.1016/j.cie.2017.01.013

[7] Information on https://en.wikipedia.org/wiki/ISO/IEC_31010

[8] Information on https://www.6sigma.us/etc/what-are-common-root-cause-analysis-rca-tools/

[9] Information on https://en.wikipedia.org/wiki/Root_cause_analysis

[10] Information on https://en.wikipedia.org/wiki/Failure_mode_and_effects_analysis

[11] Information on https://en.wikipedia.org/wiki/Multiple-criteria_decision_analysis

[12] Y. Luo, S. Huang, S. Cao, Application of improved fishbone diagram in the operation management, Industrial Engineering Journal (Guangzhou) 10(2) (2007) 138. https://doi.org/10.1016/j.jclepro.2017.10.334

[13] N.P. Suh, Designing-in of quality through axiomatic design, IEEE Trans. Reliab. 44(2) (1995) 256–264, https://doi.org/10.1109/24.387380

[14] N.P. Suh, Axiomatic design theory for systems, Res. Eng. Des. 10(4) (1998) 189–212. https://doi.org/10.1007/s001639870001

[15] N.P. Suh, Axiomatic Design: Advances and Applications, Oxford University Press, New York, 2001.

[16] S. Lu, A. Liu, Innovative design thinking for breakthrough product development, Procedia CIRP 53 (2016) 50–55. https://doi.org/10.1016/j.procir.2016.07.034

[17] M. Nordlund, S.G. Kim, D. Tate, T. Lee, H.L. Oh, Axiomatic design: making the abstract concrete, Procedia CIRP 50 (2016) 216–221.

Innovative Manufacturing Engineering and Energy - IMANEE2025   Materials Research Forum LLC
Materials Research Proceedings 61 (2026) 58-66   https://doi.org/10.21741/9781644903995-8

Practical aspects regarding optimization of three axis CNC machining

Mihail BÎCIOC[1,a*], Sergiu MAZURU[2,b]

[1]Technical University of Moldova, Department of Mechanical Engineering, Chisinau, Republic of Moldova

[2]Technical University of Moldova, Department of Industrial Design and Manufacturing, Chisinau, Republic of Moldova

[a]mihail.bicioc@doctorat.utm.md, [b]sergiu.mazuru@tcm.utm.md

Keywords: CNC Milling, Three-Axis Machining, Machining Efficiency, Toolpath, G-Code

Abstract. In CNC machining, optimizing the manufacturing process is essential for reducing costs and increasing efficiency. This article presents practical methods for optimizing G-code using CNC simulation and optimization software. The study focuses on analyzing an existing CNC program created for a Doosan CNC machine, with the primary goal of reducing machining time and optimizing tool paths.The optimization process involves identifying inefficiencies within the original G-code and implementing improvements through various strategies. The most significant modifications include adjusting the feed rate, increasing the depth of cut, adopting adaptive machining strategies, and eliminating unnecessary movements from the G-code. These changes aim to enhance machining performance while maintaining the quality and accuracy of the final product.At the final stage, the optimized data is presented, demonstrating a notable improvement. Specifically, the machining program shows a reduction in processing time by approximately three minutes compared to the initial version. This enhancement translates into increased productivity and more efficient resource utilization. Additionally, the study highlights the importance of continuous analysis and optimization to further improve CNC machining processes.

Introduction

As the requirements, quantities, and complexity of parts have increased, the need for automation in the machining process has become more evident. Unlike manual machining [1], where the operator directly controls the tool and its path, and operations are performed step by step with frequent adjustments and repositioning, the accuracy, precision, and surface quality of the parts largely depend on the operator's skill. The response to current demands comes in the form of CNC (Computer Numerical Control) machining [2], where most of the process relies on computer-generated CAM (Computer-Aided Manufacturing) programs [3], which convert 3D models into automated G-code instructions.

Modern G-code is no longer just a sequence of basic commands. It now includes optimized toolpaths, subroutines, macro commands, and even adaptively generated code in real-time, depending on machining conditions. Furthermore, modern software allows G-code optimization through analysis and simulation [4–5], reducing machining ambiguities and eliminating programming errors. This evolution has completely transformed the way CNC machines are programmed, offering unmatched flexibility and efficiency compared to non-CNC machining methods. Although traditional machining may incur lower costs, CNC machining is often superior in terms of processing time, precision, and manufacturing efficiency. In addition to its multifunctionality—combining milling, turning, and drilling—the ability to machine complex geometries and achieve non-standard precision levels is a key differentiator [6–7]. Despite these advanced features, G-code remains an open parameter for optimization of non-standard parts [8–10].

Innovative Manufacturing Engineering and Energy - IMANEE2025   Materials Research Forum LLC
Materials Research Proceedings 61 (2026) 58-66   https://doi.org/10.21741/9781644903995-8

Recently, with increasing market competition, production efficiency and cost control have become daily priorities in the manufacturing industry. CNC machining companies are focused on improving adaptability and boosting productivity by monitoring every operation, starting from personnel training and equipment maintenance to ensuring the supply chain of raw materials, tools, and machinery [11–12]. In this context, CNC process optimization has become a strategic point for development and market sustainability. As a result, companies are increasingly investing in modern G-code simulation and optimization technologies, as well as in the use of advanced cutting strategies [13].

In this context, the present paper aims to present practical methods for G-code optimization using modern CNC simulation and optimization software. The analysis focuses on the complete process of creating a workpiece, highlighting concrete strategies for improving surface quality, reducing machining time, and optimizing toolpaths. Technical aspects such as feed rate, depth of cut, the application of adaptive strategies, and the elimination of unnecessary movements from the G-code will also be discussed—all contributing to a faster, more precise, and more efficient manufacturing process [14].

Stages of G-code optimization:
- Simulation and visualization of the program
- Adjusting feed rate and depth of cut
- Applying adaptive strategies and rearranging operations
- Identifying idle movements and optimizing the toolpath

Materials and Equipment

For the analysis of the CNC program optimization process, a 3-axis Doosan DNM 5700 CNC machine tool (Fig. 1) was used, equipped with a Fanuc control system. It is capable of machining complex surfaces and achieving high precision. The technical specifications are presented in Table 1.

Figure 1: The Doosan DNM 5700 CNC machine

The design and generation of the G-code were carried out using CAM software, specifically SolidCAM, which allows for the simulation of the milling process and the adjustment of the toolpath based on the part geometry and machining parameters. The generated code was then imported into an additional simulator, namely CIMCO Edit, for further analysis and optimization.

Table 1. Technical Specifications – Doosan DNM 5700

Specification	Value
X-axis travel	1,050 [mm]
Y-axis travel	570 [mm]
Z-axis travel	510 [mm]
Table size	1,300 [mm] × 570 [mm]
Max. table load	1,000 [kg]
Max. spindle speed	12,000 [rpm]
Spindle motor power	18.5 [Kw]
Tool holder type	BT40
Tool magazine capacity	30 tools
CNC control system	FANUC i Series
Machine weight	approx. 7,000 [kg]

The part analyzed in this study (Figure 2) is made of 6061 aluminum alloy. This material is commonly used in industrial applications due to its good machinability and mechanical strength. The geometry of the part includes flat surfaces, holes, and cavities, making it suitable for testing and optimizing toolpaths.

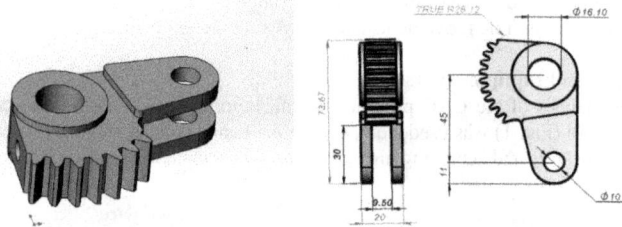

Figure 2: 3D model for testing

Case Study

As previously mentioned, the part used in this study is a prototype intended for testing a CNC machining program. Based on the analysis of the part's geometry, it mainly consists of flat surfaces, several holes, and a gear segment where milling of the gear profile is proposed.

To simplify the study and focus exclusively on the machining program, tight dimensional tolerances of the model were excluded during execution.

The idea of optimizing the CNC program arose from the need to perform a preliminary cost estimation for potential mass production of the part. The initial version of the program (Fig. 3) was analyzed, revealing several possibilities for improvement, particularly regarding the tool's positioning and retraction movements, as well as the adjustment of machining parameters such as cutting speeds and depths.

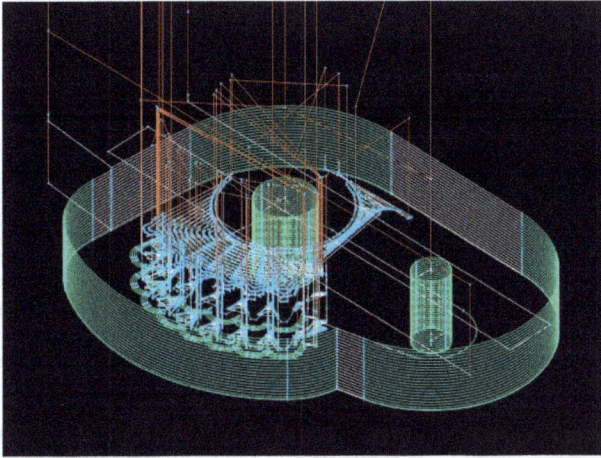

Figure 3: Initial toolpath

The optimization process began by improving the tool positioning and retraction strategy as shown in (Fig. 4) following the surface and contour machining of the first tool, but also by modifying the method of introducing the tool into the material. The standard linear entry was replaced by a tangential entry, which ensures a smoother transition and reduces mechanical stress on both the tool and the workpiece.

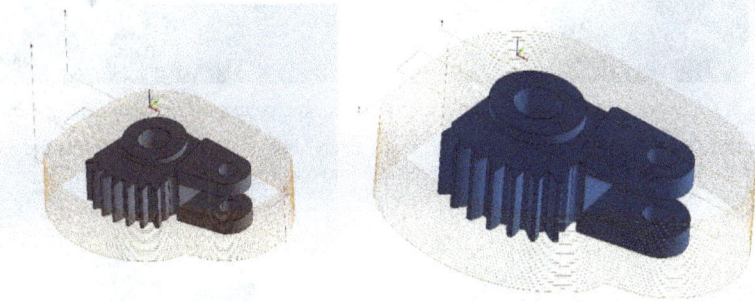

Figure 4: The image on the left represents the initial CNC program.The image on the right represents the optimized program.

The next step in optimizing machining time, in addition to reducing the positioning distance, was to optimize the tool feed rate during positioning (Fig. 5). Specifically, the rapid feed rate was increased from F300 to F1500, which contributed to shortening the total positioning time without compromising accuracy.

```
(TOOL -19- MILL DIA 8.0 R0. MM )          N3 T19 M6
G90 G00 G40 G54                           (TOOL -19- MILL DIA 8.0 R0. MM )
G43 H19 D19 T25 G0 X-4.424 Y-44.002 Z25. S3500 M3   G90 G00 G40 G54
M8                                        G43 H19 D19 T25 G0 X-4.424 Y-44.002 Z2. S3500 M3
(--------------------)                    M8
(F-CONTOUR-1 - PROFILE)                   (--------------------)
(--------------------)                    (F-CONTOUR-1 - PROFILE)
   X-4.424 Y-44.002 Z2.                   (--------------------)
G1 Z-17.2 F300                               X-4.424 Y-44.002 Z2.
G41 G1 X-0.439 Y-44.367 F850              G1 Z-17.2 F1500
```

Figure 5: *Modification of the engagement feed rate.*

Figure 6 shows the initial machining path for a depth of Z–3 mm. In this case, it was decided that, in addition to the previously mentioned changes regarding feed rates, tool entry, and clearance height between areas, the machining strategy would also be adjusted. The approach was changed from simple climb milling to a combined climb zigzag strategy, where the tool moves in both directions. This maintains a favorable cutting regime while reducing travel distances, without compromising machining quality.

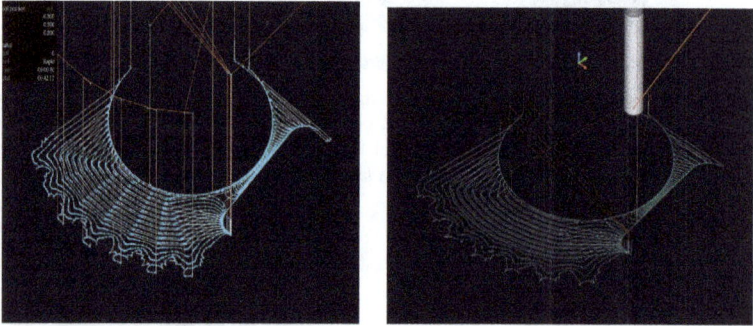

Figure 6: *Comparison of the machining path for the pocket, compared to the optimized version.*

The next step was the optimization of the work depth of the cutter, which performs the release between the tooth profile and the cutter responsible for roughing these features (Fig. 7).

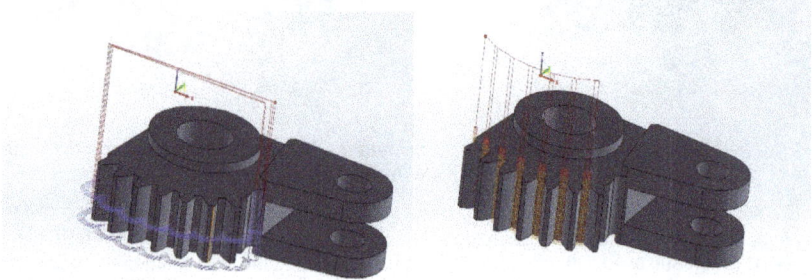

Figure 7: *Initial roughing of the gear profile.*

Subsequently, the front surface was finished, making it possible to change the starting and retraction depth of the drill after the roughing or, better said, material release stage of the tooth

Innovative Manufacturing Engineering and Energy - IMANEE2025   Materials Research Forum LLC
Materials Research Proceedings 61 (2026) 58-66   https://doi.org/10.21741/9781644903995-8

profile. First, the positioning on the Z axis was changed, moving it from the initial position of Z 25 mm to Z -2 mm. Another parameter that was changed is the R parameter, which controls the tool retraction plane (Fig. 8).

```
(*TOOL 18 - DIA 1.2-*)                          ( TOOL -18- DRILL DIA 1.2 MM )
N5 T18 M6                                        G90 G00 G40 G54
( TOOL -18- DRILL DIA 1.2 MM )                   G43 H18 D18 T25 G0 X0.771 Y-40.839 Z2. S5836 M3
G90 G00 G40 G54                                  M8
G43 H18 D18 T25 G0 X0.771 Y-40.839 Z120. S5836 M3   (---------------)
M8                                               (D-DRILL1 - DRILL)
(---------------)                                (---------------)
(D-DRILL1 - DRILL)                                  X0.771 Y-40.839 Z-2.
(---------------)                                G98 G83 Z-18. R-2.4 Q0.5 F291
   X0.771 Y-40.839 Z25.                             X5.392 Y-43.514
G98 G83 Z-18. R-0.5 Q0.5 F291                       X6.484 Y-43.997
```

Figure 8: *Comparison of the initial G-code (left) and the modified G-code (right) at the drilling stage*

The final finishing operation did not undergo major changes; only the positioning and retraction distances of the tool, as well as their feed rates, were reduced. The final toolpaths obtained after the modifications to the G-code can be seen in the figure below.

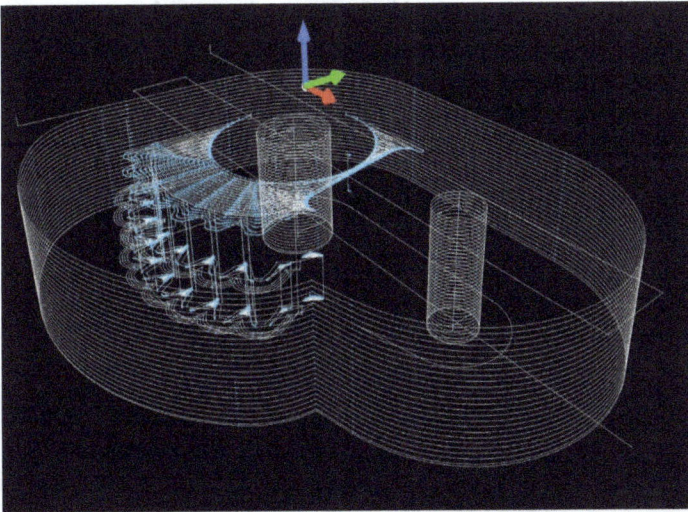

Figure 9: *Toolpath after the final modifications.*

Results

For a clear and detailed comparison, the analysis will present both the milling time data, the positioning and retraction time, as well as the total machining time. This will provide a comprehensive view of the impact of optimization on process performance. The first figure (fig.10) presents the initial machining data, where the total machining time is 36 minutes and 58 seconds. Of this, approximately 33 minutes represent intensive milling time, while approximately 3 minutes are allocated to the positioning, retraction, and transition from one surface to another.

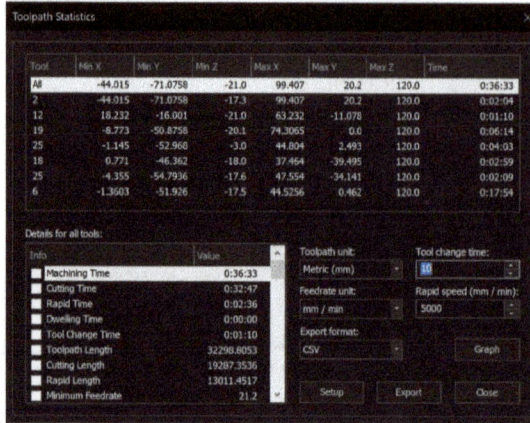

Figure 10: Initial machining time.

The second figure (fig.11) presents the final data of the program, after the optimizations were implemented. As a result, the total machining time was significantly reduced, with a gain of approximately 2 minutes for continuous machining and a gain of approximately 1 minute for retraction and repositioning between surfaces. This reduction was made possible by optimizing each toolpath, with improvements ranging from a few seconds to a minute, leading to a considerable enhancement of the overall process performance.

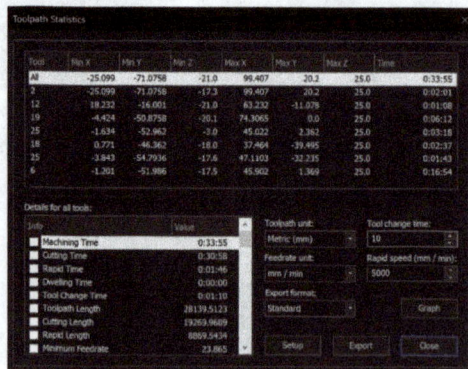

Figure 11: Optimized machining time.

Based on the total time reduction of approximately 3 minutes per part achieved by the optimization demonstrated above by optimizing the g-code, a significant economic gain can be understood when the process is applied to a batch of 100 parts, for example. Thus, the total time savings amount to approximately 300 minutes, equivalent to 5 hours of machining. Considering an average cost of 35 euros per hour for using a CNC machine in Moldova, this results in a theoretical reduction of 175 euros per batch of 100 parts. This gain reflects the increased efficiency of the optimization method and the direct benefits on the total product cost.

Innovative Manufacturing Engineering and Energy - IMANEE2025   Materials Research Forum LLC
Materials Research Proceedings 61 (2026) 58-66   https://doi.org/10.21741/9781644903995-8

Conclusions

In conclusion, we can say that CNC programs, specifically G-code optimization through modifications in machining techniques and the adjustment of positioning parameters, represent a valuable criterion for increasing the efficiency of machining processes. By reanalyzing and adjusting toolpaths, positioning movements, and cutting parameters, significant reductions in milling time can be achieved, directly impacting machining costs.

This study has demonstrated that, through the interventions made to the program, affordable prices can be generated, competitive with larger companies. In an industrial context moving towards a sustainable economy, the adoption of G-code optimization practices is no longer just an option, but a strategic necessity to ensure production sustainability and profitability.

To validate the generated program, the final part produced (Fig.12) for testing purposes is presented below, with no programming errors or unexpected collisions identified during the machining process.

For the future, I would propose introducing an additional checkpoint for the CNC program before the actual machining process. This intermediate step could include a detailed check of the G-code, using both additional simulations and manual analysis to identify and eliminate errors and reduce machining time.

Implementing such a checkpoint would increase the reliability of the production process, preventing unforeseen situations and reducing the risk of collisions or other failures. Moreover, this measure would ensure more thorough preparation before the actual production, leading to better time and resource management in the workshop.

Figure 12: Executed model.

References

[1] Frățilă, D., Radu, A., Păcurar, A., Păcurar, R., Conțiu, G., Panc, N., & Pop, G., Tehnologii de fabricație, 2020.

[2] But, A., & Mazuru, S., "Serghei Scaticailov: Fabricația asistată de calculator," Universitatea Tehnică a Moldovei, Chișinău: Tehnica-UTM, 2021.

[3] But, A., & Scaticailov, S., "Using CAM software to improve productivity," IOP Conference Series: Materials Science and Engineering, vol. 564, no. 1, p. 012055, 2019. https://doi.org/10.1088/1757-899X/564/1/012055

Materials Research Forum LLC
https://doi.org/10.21741/9781644903995-8

[4] Lysek, K., Gwiazda, A., & Herbus, K., "Application of CAM systems to simulate milling machine work," IOP Conference Series: Materials Science and Engineering, vol. 400, no. 4, p. 042037, 2018. https://doi.org/10.1088/1757-899X/400/4/042037

[5] Dodok, T., Čuboňová, N., Císar, M., Kuric, I., & Zajačko, I., "Utilization of strategies to generate and optimize machining sequences in CAD/CAM," Procedia Engineering, vol. 192, pp. 113-118, 2017. https://doi.org/10.1016/j.proeng.2017.06.078

[6] Bîcioc, M., "Technological assurance of mechanical processing of complex shaped parts based on machining centers with numerical control," Journal of Engineering Sciences, no. 3, pp. 8-17, 2024.

[7] Bîcioc, M., "A case study of the tool paths strategies applicable for milling complex surfaces on three coordinates CNC machines," Acta Technica Napocensis-Series: Applied Mathematics, Mechanics, and Engineering, pp. 505-512, 2024. https://doi.org/10.24136/atna.2024.212

[8] Mazuru S., Scaticailov S. New technological solution for manufacturing precessional gears with non-standard profile. Acta Technica Napocensis. Series: Applied Mathematics, Mechanics, and Engineering. Vol. 66, No 5, October, 2023. https://doi.org/10.24136/atna.2024.212.

[9] Bostan I., Mazuru S., Scaticailov S. Processes generating non-standard profiles variable convex- concav of precessional gear. Journal of Engineering Sciences and Innovation. Volume 5, Issue 2 / 2020.

[10] Bostan I., Mazuru S., Scaticailov S. Technologies for precessional planetary transmissions toothing generation. TEHNOMUS jurnal. Nr. 20. 2013

[11] Botnari Vl., Mazuru S. Influence of Processing Parameters on the Quality of the Superficial Layer after Processing Surfaces with Plastic Deformation Processes. Applied Mechanics and Materials Vol. 657. 2014.

[12] Stroncea A., Toca A., Nitulenco T. Considerations regarding the optimal dimensional design of machining technologies. Acta Technica Napocensis. Series: Applied Mathematics, Mechanics, and Engineering. Vol. 66, Issue Special II, October, 2023.

[13] Toca A., Stroncea A., Stingaci I., Rusica I. Synthesis of optimal dimensional structure of the technological processes of machining. IOP Conf. Ser.: Materials Science and Engineering 400, United Kingdom, 022054. 2018.

[14] Toca A., Iațchevici V., Nițulenco T., Rusu N. Some aspects of technology transfer. MATEC Web of Conferences 178, France, 08006.2018.

Innovative Manufacturing Engineering and Energy - IMANEE2025
Materials Research Proceedings 61 (2026) 67-76

Materials Research Forum LLC
https://doi.org/10.21741/9781644903995-9

Research on the use of advanced imaging techniques while respecting intellectual property protection

Dan Theodor ANDRONIC[1,a], Aurel Mihail TITU[2,b] *

[1]Doctoral Candidate, National University of Science and Technology Politehnica Bucharest, Faculty of Industrial Engineering and Robotics, Bucharest, Romania

[2]Professor, Corresponding Author, "Lucian Blaga" University of Sibiu, 10 Victoriei Street, Sibiu, Romania

[a]andronic_danth@yahoo.com, [b]mihail.titu@ulbsibiu.ro

Keywords: Neurosurgery, Advanced Imaging Techniques, Intellectual Property Protection, Surgical Planning, Medical Innovation, Clinical Outcomes, Diagnostic Processes, Patient Management

Abstract. The concept of input-output correlation refers to the relationship between information obtained through advanced imaging techniques (input) and clinical outcomes observed after surgical interventions (output). It is an essential component of neurosurgery, with a significant impact on diagnostic processes, therapeutic planning, and patient management. This correlation facilitates a better understanding of brain pathology and contributes to optimising surgical interventions, reducing risks, and improving patient outcomes. This model is based on data analysis and its interpretation in the context of the patient's health status. In neurosurgery, images obtained with techniques such as MRI (Magnetic Resonance Imaging) or CT (Computed Tomography) are essential for identifying structural abnormalities and assessing patient condition before surgery. The processes of interpreting brain images can include, in addition to identifying lesions, evaluating their impact on brain functions. The neurosurgeon can use MRI data to determine the tumour's precise location and assess its relationship to adjacent anatomical structures, such as blood vessels and healthy functional brain tissue. This assessment helps plan the surgical intervention, ensuring favourable patient outcomes.

1. Introduction

The Importance of Correlation in Establishing a Diagnosis

Input-output correlation plays a crucial role in establishing an accurate diagnosis. In the context of neurosurgery, precise diagnosis is essential for planning surgical interventions and the subsequent management of patients. When following clear protocols, images obtained with advanced imaging techniques provide detailed information about brain structures. As neurosurgeons, correlating this data with clinical symptoms and medical history allows us to issue precise diagnoses and therapeutic recommendations [1]. Using MRI to assess the size, type, and location of a brain tumour enables us not only to determine whether surgery is necessary but also to plan the optimal surgical approach. Tumours near critical brain structures, such as the brainstem or motor cortex, require a different approach than those in safer areas. This correlation between imaging data and intervention outcomes also helps establish patient prognosis [2,3].

Surgical Planning

Surgical planning is a crucial step in the clinical process, and the correlation between imaging data obtained by strictly adhering to imaging protocols and surgical outcomes is a key factor. Effective planning is based on a detailed evaluation of the images obtained, which provides us with a thorough understanding of brain anatomy and how lesions interact with adjacent structures [4,5].

Innovative Manufacturing Engineering and Energy - IMANEE2025   Materials Research Forum LLC
Materials Research Proceedings 61 (2026) 67-76   https://doi.org/10.21741/9781644903995-9

Advanced technology, such as functional MRI (fMRI) (Fig. 1), adds a new dimension to surgical planning. fMRI can identify areas responsible for cognitive, motor, or sensory functions, which is crucial in cases where there is a risk of impairing these functions during surgery. In treating drug-resistant epilepsy [3], we can use fMRI to map cortical areas involved in seizure generation and determine whether they can be excised without affecting essential cortical functions. Thus, correlating images with cortical functions is vital for surgical decision-making [6].

Use of Advanced Imaging Techniques

Advanced imaging techniques have revolutionised neurosurgery and significantly impacted the diagnosis and surgical planning for patients with neurological conditions [3,4]. Magnetic resonance imaging (MRI), computed tomography (CT) (Fig. 2), and molecular imaging are predominantly used in this context. Each technique has distinct characteristics, advantages, and disadvantages that influence the selection of the appropriate method based on clinical needs [2,7].

2. Types of advances in imaging techniques

Magnetic Resonance Imaging (MRI)

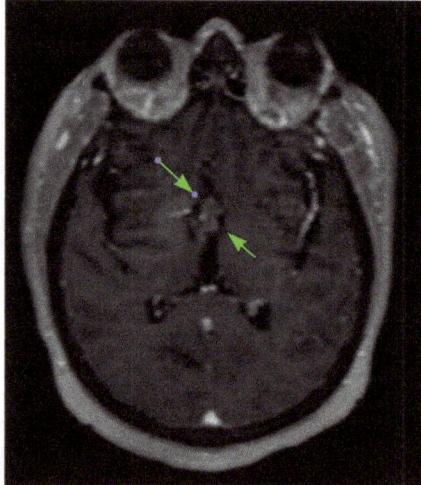

Figure 1. MRI Exploration of a patient with a brain tumour. (Personal contribution).

MRI is a non-invasive imaging technique that utilises magnetic fields and radio frequencies to create detailed images of the internal structures of the human body. This technique is especially valuable for evaluating brain pathologies because it offers images in multiple anatomical planes [8,9].

Table 1. Advantages and disadvantages of MRI exploration [1].

Advantages of MRI		Disadvantages of MRI	
High Resolution	MRI provides excellent soft-tissue resolution, allowing detailed identification of brain structures, including grey matter, white matter, and blood vessels.	Long examination time	MRI can require significant time to obtain images, which can be problematic in some instances.
No Ionising Radiation	Unlike CT, MRI does not involve exposure to ionising radiation, making it a preferable choice among young patients and pregnant women.	Limitations in the case of claustrophobic patients	Some patients may develop panic attacks due to the enclosed space of the MRI machine, which can complicate the diagnostic process.
Functional Images	Using functional MRI (fMRI), it is possible to monitor brain activity in real time, providing detailed information about the areas activated during different cognitive tasks.	Reduced sensitivity to calcifications	MRI may have difficulty detecting calcifications.

Computed Tomography (CT)

CT is an imaging technique that utilises X-rays to produce cross-sectional images of the body. This rapid method provides detailed images, making it particularly beneficial in emergencies [10].

Table 2. Advantages and disadvantages of CT exploration [5].

Advantages of CT:		Disadvantages of CT:	
Rapidity	CT can obtain images in a very short time, making it ideal for evaluating patients with head trauma or haemorrhage strokes.	Radiation exposure	The procedure involves ionising radiation, which can be a disadvantage with repeated exposure, especially in young patients.
Effective detection haemorrhage's	CT is the method of choice of for the rapid identification of intracranial haemorrhages.	Lower resolution than soft tissue MRI	Although CT is effective for evaluating bony structures, it has lower soft-tissue resolution than MRI.
Visibility of bone structures	CT provides clear images of bone structures and can detect fractures with high accuracy.	Limitations in diagnosing neurological conditions	It is less effective in detecting subtle brain conditions, such as micro-infarcts or anatomical abnormalities.

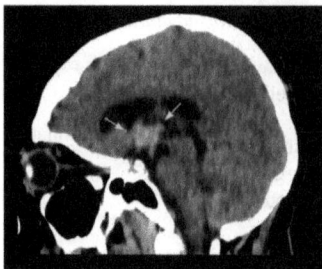

Figure 2. CT Exploration of the patient with a brain tumour (Personal Contribution).

Molecular Imaging

Molecular imaging is a relatively new frontier in imaging, capable of visualising and characterising biological processes at the cellular and molecular levels. This technique utilises radioisotopes or molecular markers to highlight specific metabolic activity in tissues, providing valuable insight into the pathological state of the brain [11,12].

Table 3. Advantages and disadvantages of molecular imaging [4].

Advantages of molecular imaging:		Disadvantages of molecular imaging:	
Early detection of pathologies	Molecular imaging can detect subtle changes, enabling early diagnosis of neurological conditions.	High costs	Molecular imaging procedures are often more expensive compared to MRI and CT, which may limit affordability for some patients.
Tumour characterization	This technique allows the characterisation of tumours based on their metabolism and biological activity, helping distinguish malignant from benign tumours and assess tumour aggressiveness.	Limited availability	Not all hospitals and medical centres have the necessary equipment for molecular imaging, making it not always a viable option.
Treatment personalization	By highlighting tumour responses to different therapies, molecular imaging can contribute to the adoption of personalised therapeutic regimens, thereby maximising treatment effectiveness.	The need for an interpretation specialist	Interpretation of images obtained through molecular imaging requires a high level of specialisation, and the availability of these experts is limited.

3. Comparing the advantages and disadvantages of each technique

When selecting the most appropriate imaging technique, it is crucial to weigh each method's advantages and disadvantages based on clinical needs and circumstances. In emergencies (Fig. 3), CT is preferred for its speed, while MRI may provide a more detailed assessment of brain structures (Table 1). For evaluating brain tumours, molecular imaging could be utilised to assess specific biological and metabolic characteristics; however, physicians need to be aware of the available resources and the expertise required to interpret the results (Table 2) [1,2].

Innovative Manufacturing Engineering and Energy - IMANEE2025   Materials Research Forum LLC
Materials Research Proceedings 61 (2026) 67-76   https://doi.org/10.21741/9781644903995-9

Integrating advanced imaging techniques into a multidisciplinary approach is essential for optimising the management of patients with neurological conditions. At the same time, adhering to clear radiological exploration protocols can streamline collaboration between neurosurgeons, radiologists, and oncologists. Interdisciplinary collaboration, supported by these techniques and adherence to examination protocols, enables a complex and detailed assessment that supports well-founded therapeutic decisions (Table 3) [13,14].

Therapeutic protocols can be developed by correlating data from various imaging sources. A brain tumour should first be evaluated with a CT scan to quickly detect haemorrhage when necessary and subsequently assessed by a contrast-enhanced MRI to obtain detailed information about its size and anatomical impact [1].

The medical intervention process does not end with the original images [15]; instead, postoperative images also play an essential role in evaluating treatment effectiveness and potential complications. Therefore, an integrated approach employing various imaging techniques throughout the patient care is crucial [3,4].

Advancing techniques, including MRI, CT, and molecular imaging, form the foundation for evaluating and managing neuro-oncological diseases. Each method has advantages and disadvantages [4], and the appropriate technique depends on the specific clinical context and the diagnostic assumptions that must be addressed. Using these methods, integrated through transparent and collaborative protocols, improves prognosis [6,7].

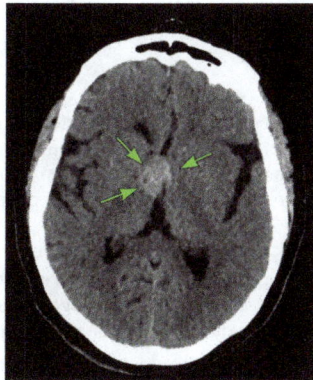

Figure 3. CT of the same patient with the brain tumour, but a different scanning protocol (Personal contribution).

4. Importance in neurosurgery

Neurosurgery is constantly evolving, influenced by technological and scientific advances that have reshaped how neurological conditions requiring surgery are diagnosed and treated. Advanced technologies are essential in this speciality, facilitating more precise, safe, and effective interventions. In this paper, we aim to discuss the evolution of modern neurosurgery, highlighting the impact of advanced technologies on clinical practice and the factors that influence surgical decisions based on the available imaging [3,16].

Significant advances in imaging technology, surgical specialisations, and biomedical research have marked the evolution of neurosurgery. Since its beginnings, when neurosurgery relied on rudimentary techniques and clinical observations, the field has evolved rapidly, integrating advanced technologies to improve the diagnosis and treatment of brain diseases [17]. One of the most critical steps in this evolution was the introduction of MRI in the 1980s, which revolutionised medical imaging. This technology enabled doctors to obtain detailed brain images without exposing patients to ionising radiation, a significant issue with computed tomography (CT). MRI has dramatically enhanced the ability to diagnose tumours, brain lesions, and congenital malformations, facilitating more precise surgical interventions [1,18].

Additionally, advances in molecular imaging technologies, such as PET (Positron Emission Tomography) and SPECT (Single Photon Emission Tomography), have introduced a new level of sophistication to assessing the metabolic state of brain tissue. These techniques aid in diagnosing tumours and facilitate monitoring treatment response and disease progression, thereby enabling a personalised approach to patient care [19,20].

Factors Influencing Surgical Decisions

Surgical decisions in neurosurgery are often influenced by factors such as the availability of imaging technologies, the complexity of the condition, and the patient's overall health [21].

Availability of Imaging Technologies

Access to advanced imaging technologies is key to formulating surgical strategies. In medical centres with high-resolution MRI and CT, neurosurgeons can conduct a more detailed assessment of brain lesions, providing a solid basis for therapeutic decisions. Additionally, intraoperative imaging technologies provide continuous updates during the intervention, facilitating quick reactions and adaptations to the surgical strategy based on real-time observations [22,23]. Furthermore, adherence to imaging investigation protocols ensures a uniform assessment of patients, allowing all neurosurgeons access to the same levels of information and diagnosis. This reduces diagnosis variability and helps achieve more consistent results in surgical interventions [3].

Complexity of the Condition

The complexity of the neurological condition is another determining factor in surgical decisions. In the case of brain tumours, the size, type, and location of the tumour, along with the involvement of adjacent structures, can significantly influence the decision to intervene. Tumours in sensitive brain areas (Fig. 4) necessitate more careful evaluation and planning, often involving imaging techniques to help minimise risks [2].

On the other hand, conditions like treatment-resistant epilepsy may necessitate specific neurosurgical interventions, such as resecting certain sections of the cerebral cortex. These interventions are typically planned based on extensive image evaluations showing the location of the epileptogenic focus; so, the importance of advanced imaging in setting an appropriate operational plan is underscored [2].

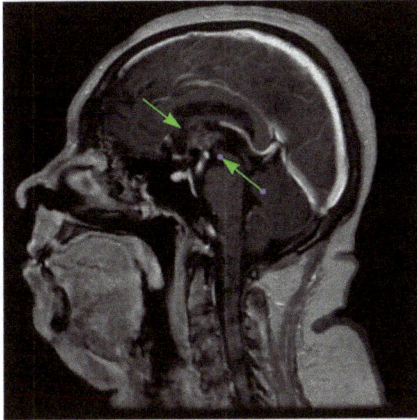

Figure 4. Brain tumour evaluation by MRI exploration (Personal contribution).

Patient General Condition

The patient's overall condition is another crucial factor that influences surgical decisions. Evaluating risk factors, such as comorbidities and the patient's age, is essential. An elderly patient with multiple comorbidities may face a higher risk of postoperative complications, and surgical decisions will take these factors into account [4].

Advanced imaging technologies can help assess the patient's overall condition by providing detailed images of affected brain structures and the impact of lesions on neuronal function. This allows us to evaluate the lesion and its interaction with the patient's health [1].

5. Compliance with intellectual property protection criteria

Intellectual property (IP) is a legal concept that protects the results of individuals' or entities' creative and intellectual activities. It encompasses a set of legal rights that enable the recognition and protection of inventions, literary and artistic works, trademarks, and industrial designs. The significance of intellectual property is particularly pronounced in the technological and medical fields, as it plays a crucial role in stimulating innovation and safeguarding the commercial interests of researchers, technology developers, and companies [24].

In the technological field, protecting intellectual property is vital for stimulating innovation. Companies invest significant amounts in research and development, and safeguarding ideas, inventions, or technical details through patents provides them with a competitive advantage. Intellectual property rights offer investors and creators legal protection that enables them to control the use of their innovations, ensuring a recovery of their investment costs and opportunities for monetisation [1].

Intellectual property plays a crucial role. Inventions and discoveries in medicine, treatments, or medical technologies often result from complex and costly research. Owning intellectual property rights for these innovations provides economic incentives for research, contributing to the development of new therapies and advancements in disease treatment [25].

The Importance of Intellectual Property Protection in Medical Innovation

Intellectual property plays a crucial role in supporting medical and technological innovation. It protects inventions and discoveries and encourages the research and development of new treatments and technologies. Without adequate intellectual property protection, many companies would be unwilling to invest in the expensive, risky research required to develop new products [1].

Another controversial issue is the "patents versus innovation" dilemma. It is argued that a highly restrictive patent system can inhibit further innovation, as researchers face difficulties developing new products based on original research. This raises questions about the need to reward patent management in ways that encourage collaboration and minimise bottlenecks to innovation [2].

6. Conclusion

The evolution of neurosurgery has been driven by fascinating advances in imaging technology and continuous innovation in the field. Advanced technologies such as MRI, CT, and molecular imaging have transformed clinical practice, enabling neurosurgeons to achieve unprecedented precision in diagnosing and treating brain disorders [3].

The factors influencing surgical decisions are variable and complex, encompassing the availability of imaging technologies, the complexity of the neurological condition, and the patient's overall health. This interrelationship underscores the importance of advanced imaging in neurosurgery, highlighting that therapeutic decisions today rely not only on clinical symptoms but also on a well-defined informatics framework, which furnishes essential information for effective and safe surgical intervention [2]. Thus, utilising advanced technologies enhances patient outcomes and contributes to the ongoing development of high-quality standards in neurosurgery [1].

Intellectual property is the backbone of innovation in the technological and medical fields. By ensuring adequate protection for inventions, works, and designs, intellectual property protection encourages investment in research and development, stimulating scientific progress and advancement. Although significant challenges and controversies exist in this area, relevant legislative frameworks and institutions must collaborate on developing an equitable system that supports innovation while ensuring the accessibility of treatments for all patients [25].

As we continue to navigate technological evolution and scientific discoveries, protecting intellectual property will remain a top priority, profoundly shaping the future of global health and human progress. All stakeholders—from researchers and pharmaceutical companies to legislators and patients—must recognise the significance of intellectual property and collaborate to create a system that promotes both innovation and equitable access to medical resources [1,26].

Intellectual property serves as a critical driver of technological and medical progress. In a world where rapid innovation is increasingly vital, ensuring adequate protection of this property has become a crucial issue for the future of human health. It remains essential for all stakeholders to collaborate in creating a system that supports innovation and equitable access to medical resources for all patients worldwide [25].

Protecting intellectual property is a legal and economic necessity, as well as a moral responsibility, to ensure that scientific and medical progress benefits society equitably. Only through an integrated and balanced approach to intellectual property can we move towards a future in which innovation and accessibility complement each other, maximising benefits for patients and society [27].

References

[1]   D. Andronic, A. M. Țîțu, and A.-F. Iamandii, "Management of exploration of traumatic injuries through protocols in the field of neurosurgery and the importance of intellectual property in this context." Journal of Research and Innovation for Sustainable Society, vol. 6, no. 2, p. 7, Nov. 2024, doi: 10.33727/jriss.2024.2.1:7-17.

[2]   A. C. Tan, D. M. Ashley, G. Y. López, M. D. Malinzak, H. S. Friedman, and M. Khasraw, "Management of glioblastoma: State of the art and future directions." CA A Cancer Journal for Clinicians, vol. 70, no. 4. Wiley, p. 299, Jun. 01, 2020. doi: 10.3322/caac.21613.

Materials Research Forum LLC
https://doi.org/10.21741/9781644903995-9

[3] N. H. Vadhavekar et al., "Advancements in Imaging and Neurosurgical Techniques for Brain Tumor Resection: A Comprehensive Review." Cureus, vol. 16, no. 10. Cureus, Inc., Oct. 31, 2024. doi: 10.7759/cureus.72745.

[4] D. Andronic and A. M. Țîțu, "Needed To Save A Life, Point Of View From Experience." Annals of the Academy of Romanian Scientists Series on Economy Law and Sociology, vol. 7, no. 1, p. 10, Jan. 2024, doi: 10.56082/annalsarscieco.2024.1.10.

[5] D. T. Andronic, O. R. Chivu, A. Semenescu, D. F. Marcu, and A. M. Țîțu, "Nonconventional Radiological Technologies And Processes Used For Neurosurgery In Correlation With Intellectual Property Protection With Direct Application In Healthcare Organization." Jun. 2025. [Online].
Available: https://www.revtn.ro/index.php/revtn/article/view/522

[6] S. Mahboob and S. Eljamel, "Intraoperative image-guided surgery in neuro-oncology with specific focus on high-grade gliomas." Future Oncology, vol. 13, no. 26. Future Medicine, p. 2349, Nov. 01, 2017. doi: 10.2217/fon-2017-0195.

[7] J. Y. Nam and J. de Groot, "Treatment of Glioblastoma." Journal of Oncology Practice, vol. 13, no. 10. American Society of Clinical Oncology, p. 629, Oct. 01, 2017. doi: 10.1200/jop.2017.025536.

[8] N. M. Ghadi and N. H. Salman, "Deep Learning-Based Segmentation and Classification Techniques for Brain Tumor MRI: A Review," Journal of Engineering, vol. 28, no. 12. Chulalongkorn University, p. 93, Dec. 01, 2022. doi: 10.31026/j.eng.2022.12.07.

[9] P. T. Krishnan, K. Pradeep, M. Khandelwal, D. Gupta, A. Nihaal, and T. S. Kumar, "Enhancing brain tumor detection in MRI with a rotation invariant Vision Transformer." Frontiers in Neuroinformatics, vol. 18, Jun. 2024, doi: 10.3389/fninf.2024.1414925.

[10] Z. Ali, A. Hamdoun, and A. Alattaya, "Advances in Radiological Techniques for Cancer Diagnosis: A Narrative Review of Current Technologies." Deleted Journal, vol. 2, no. 1. p. 43, Jan. 01, 2024. doi: 10.61838/kman.hn.2.1.6.

[11] R. Harada, S. Furumoto, Y. Kudo, K. Yanai, V. L. Villemagne, and N. Okamura, "Imaging of Reactive Astrogliosis by Positron Emission Tomography." Frontiers in Neuroscience, vol. 16. Frontiers Media, p. 807435, Feb. 08, 2022. doi: 10.3389/fnins.2022.807435.

[12] M. Beaurain et al., "Innovative Molecular Imaging for Clinical Research, Therapeutic Stratification, and Nosography in Neuroscience." Frontiers in Medicine, vol. 6. Frontiers Media, Nov. 27, 2019. doi: 10.3389/fmed.2019.00268.

[13] K.-M. Schebesch et al., "Clinical Benefits of Combining Different Visualization Modalities in Neurosurgery." Frontiers in Surgery, vol. 6, Oct. 2019, doi: 10.3389/fsurg.2019.00056.

[14] S. A. Abidi et al., "Using Brain Tumor MRI Structured Reporting to Quantify the Impact of Imaging on Brain Tumor Boards." Tomography, vol. 9, no. 2, p. 859, Apr. 2023, doi: 10.3390/tomography9020070.

[15] E. Çivelek, Y. Akyuva, N. Kaplan, F. Diren, O. Boyalı, and S. Kabataş, "Challenges in the clinical and radiological differential diagnosis of cerebrovascular events and malignant primary brain tumors: reports from a retrospective case series." Turkish Neurosurgery, Jan. 2020, doi: 10.5137/1019-5149.jtn.30371-20.2.

[16] M. D. Richardson, J. J. Parker, and A. Waziri, "Advances in neurosurgery." Neurology Clinical Practice, vol. 2, no. 3, p. 201, Sep. 2012, doi: 10.1212/cpj.0b013e31826af22b.

Innovative Manufacturing Engineering and Energy - IMANEE2025   Materials Research Forum LLC
Materials Research Proceedings 61 (2026) 67-76   https://doi.org/10.21741/9781644903995-9

[17] E. Ogando-Rivas et al., "Evolution and Revolution of Imaging Technologies in Neurosurgery." Neurologia medico-chirurgica, vol. 62, no. 12, p. 542, Oct. 2022, doi: 10.2176/jns-nmc.2022-0116.

[18] D. Larrivee, New Advances in Magnetic Resonance Imaging. IntechOpen, 2023. doi: 10.5772/intechopen.111259.

[19] W. B. Overcast et al., "Advanced imaging techniques for neuro-oncologic tumor diagnosis, with an emphasis on PET-MRI imaging of malignant brain tumors." Current Oncology Reports, vol. 23, no. 3. Springer Science+Business Media, p. 34, Feb. 18, 2021. doi: 10.1007/s11912-021-01020-2.

[20] H. W. Schroeder and L. T. Hall, "Molecular Imaging of Brain Metastases with PET." in Exon Publications eBooks, 2022, p. 1. doi: 10.36255/exon-publications.metastasis.brain-metastases.

[21] K. L. Cole, M. C. Findlay, M. Kundu, C. M. Johansen, C. Rawanduzy, and B. Lucke–Wold, "The Role of Advanced Imaging in Neurosurgical Diagnosis." Journal of Modern Medical Imaging, Feb. 2023, doi: 10.53964/jmmi.2023002.

[22] V. Patel and V. Chavda, "Intraoperative glioblastoma surgery-current challenges and clinical trials: An update." Cancer Pathogenesis and Therapy, vol. 2, no. 4. Elsevier BV, p. 256, Dec. 02, 2023. doi: 10.1016/j.cpt.2023.11.006.

[23] T. Gárzón-Muvdi, C. Kut, X. Li, and K. L. Chaichana, "Intraoperative imaging techniques for glioma surgery." Future Oncology, vol. 13, no. 19. Future Medicine, p. 1731, Aug. 01, 2017. doi: 10.2217/fon-2017-0092.

[24] D. Andronic and A. M. Țîțu, "Management Of Imaging Evaluation Criteria In Pediatric Cases In Conjunction With Intellectual Property Protection." Annals of the Academy of Romanian Scientists Series on Economy Law and Sociology, vol. 8, no. 3, p. 4, Jan. 2025, doi: 10.56082/annalsarscieco.2025.3.4.

[25] D. Andronic and A. M. Țîțu, "Intellectual Property Protection in Modern Medical Services: Neurosurgery Versus Fine Mechanics." in Lecture notes in networks and systems, Springer International Publishing, 2025, p. 335. doi: 10.1007/978-3-031-95194-7_33.

[26] D. Andronic, A. M. Țîțu, and F. A. Iamandii, "Quality Management Implementation Systems Using The Principles Of Intellectual Property In Modern Health Systems." Review of Management and Economic Engineering, vol. 23, no. 4, p. 264, Dec. 2024, doi: 10.71235/rmee.186.

[27] M. A. Țîțu and D. Andronic, "Intellectual Property Applied To Medical Imaging Systems In Present Context." Annals of the Academy of Romanian Scientists Series on Economy Law and Sociology, vol. 6, no. 2, p. 22, Dec. 2023, doi: 10.56082/annalsarscieco.2023.2.22.

Innovative Manufacturing Engineering and Energy - IMANEE2025   Materials Research Forum LLC
Materials Research Proceedings 61 (2026) 77-83   https://doi.org/10.21741/9781644903995-10

Methodological framework for evaluating the generative capabilities of AI assisted CAD parametric modeling for AM

Nicolae Răzvan MITITELU[1,a] *, Mihaela NICOLAU[1,b], Alexandru Ionuț IRIMIA[1,c], and Cosmin-Gabriel GRĂDINARU[1,d]

[1]Department of Machine Manufacturing Technology, "Gheorghe Asachi" Technical University of Iași, Romania

[a]nicolae-razvan.mititelu@student.tuiasi.ro, [b]mihaela.nicolau@student.tuiasi.ro,
[c]alexandru-ionut.irimia@student.tuiasi.ro, [d]cosmin-gabriel.gradinaru@student.tuiasi.ro

Keywords: Artificial Intelligence, 3D Design, Generative Design, AI CAD Design

Abstract. This study proposes a theoretical methodology for the analysis of the performance of artificial intelligence systems in the automated generation of parametric CAD models, focusing on additive manufacturing. In addition, it provides a steps-based method that includes the formulation of metrics, testing, statistical analysis, error mapping, and formulation of recommendations. This method also provides a composite risk indicator that is able to provide the quantification of limitations, as well as support the formulation of decisions in industry. This method is scalable, portable, and can be applied for other applications such as topological optimization and generative design. Apart from its applicability, this research provides the first theoretical framework within which the standardization of artificial intelligence processes within CAD models can be achieved.

Introduction

Artificial Intelligence (AI) is a paradigm shift within the realm of engineering design that impacts the use of Computer-Aided Design (CAD) tools in additive design. Recent breakthroughs in the use of neural networks, particularly through generative designs, widen the application of AI design tools beyond mere automated functions. Indeed, the ability of such models, geometric models that can optimize multiple aspects such as functionality, weight, cost, and aesthetics, would mean that entire design cycles will be accelerated along with associated costs. Against this background, the use of AI as a design tool raises pertinent issues with regard to its reliability, predictability, and adaptability.

AI-based models for CAD systems provide a stepping stone for more complex design, where designs will not only be validated but will be generated through the use of AI. This will mean a shift from rule-based design to emergent design. With this technology, AI will not be assisting in design, but will provide solutions that fit specific requirements. AI technology will provide solutions that will be generated based on insights from previous designs, with multiple design requirements. Geometries that would otherwise be difficult or impractical will be generated.

Although several applications of AI within CAD and additive technologies have been dealt with extensively in recent literature, the key aspect which still seems unattended is the capacity of such systems for producing valid, consistent, and adaptive models within any engineering application. This paper proposes a comprehensive theoretical methodological framework that would facilitate the critical, systematic, and repeatable assessment of the predictive capabilities of such AI systems, engaged in the automatic generation of parametric models. Moreover, this would provide a succinct theoretical background for the subsequent standardization as well as optimization of such systems.

Foundation for predictive evaluation in AI-assisted CAD

Predictive capacity of AI systems: Also termed as predictive modeling, this refers to the capability algorithms possess, whereby they provide 3D models that accurately correspond with design considerations and changes. This includes not only accuracy for models, but also robustness with regard to semantic uncertainties, flexibility with regard to different scenarios, as well as consistency within solutions through time. In this regard, the lack of such aspects may result in many production mistakes, multiple designs, as well as production losses, particularly within industries requiring high precision with regard to their production, such as aviation or biomedical industries.

In the absence of effective methodological development, the assessment of predictive accuracy still tends to be empirical in nature, relying essentially on qualitative analyses that lack systemic verification, as highlighted in the views presented by Géron [2] and Russell [3]. Consequently, there is a fundamental necessity for a methodological approach that enables, on the one hand, the assessment of deviation of generated models with respect to initial specifications, as well as the capability of algorithms on the other hand, of generalization within different frameworks. Moreover, it is required that this methodological approach enables assessment of the degree of geometric coherence necessary within complex models that might incorporate topological dependencies as well as different engineering restrictions. Finally, this methodological approach must facilitate the assessment of adaptability of AI systems with respect to different industry-specific requirements, thus providing assurance with regard to the validity and robustness of presented solutions. Such a methodological premise is particularly necessary within the framework of incorporating AI solutions within design processes that remain immediately linked with manufacturing, as design faults immediately correspond within physical form to product defects. In addition, this methodological premise would facilitate the establishment of AI system auditing metrics that might be incorporated within global engineering quality standards.

An assessment of the predictive capabilities of AI technology in CAD, as compiled in a synthetic overview of Zhao et al. [4], requires consideration of the following aspects: coherence in terms of geometries and topology, accuracy with regard to scale, robustness of algorithms, as well as their ability to generalize. Each of the above aspects is relevant to a specific aspect of the reliability of models generated with the use of technologies such as AI, with tools required for their assessment. Creation of shared benchmarking sets for comparison of solutions such as AI in industry would also be important.

Apart from the three main dimensions, a fourth additional dimension can be considered, that of computational efficiency. Indeed, within the industrial world, the time required for generation, as well as the resource intensity involved in executing algorithms, increasingly turns into important variables. Thus, the assessment of predictive capabilities must also take into consideration the compromise between quality of modeling, as well as processing speed, alongside the energy cost involved in executing the algorithms.

Proposed methodological structure

The proposed methodology encompasses a set of structured steps supported within scientific literature, as well as adapted to meet the requirements of AI-based design within additive manufacturing. Therefore, this proposal is a tool that brings relevant elements within the professionalization process of the use of AI within engineering design, as opposed to other proposals found within scientific literature that provide empirical or qualitative analyses that might usually be required. Indeed, this proposal describes a system that encompasses the following methodological steps:

Innovative Manufacturing Engineering and Energy - IMANEE2025   Materials Research Forum LLC
Materials Research Proceedings 61 (2026) 77-83   https://doi.org/10.21741/9781644903995-10

Formal definition of evaluation parameters

At this stage, a rigorous framework for engineering validation will be established. Well-established metric measures will be identified, such as RMSE (Root Mean Square Error), which represents the mean quadratic deviation between calculated and ground truth geometric value pairs, MAE (Mean Absolute Error), representing the mean absolute deviation of geometric predictions, as well as the F1-score, modified for topological correctness with regard to geometric object classes (for example, edges, continuous surfaces, closed volumes).

To adapt the F1-score to CAD topology, each geometric entity is treated as an element of a graph structure. Let E_{gt} denote the set of ground-truth edges (or surfaces/volumes) and E_{pred} the set of predicted entities, with adjacency relations represented in the matrices A_{gt} and A_{pred}. A True Positive (TP) is defined as an entity that appears in both E_{gt} and E_{pred} and whose local adjacency neighbourhood matches the ground truth within a tolerance τ, ensuring consistency between the predicted and actual topological relationships.

Model interpretability gets measured by way of explainability tools. These tools include attention-map visualization. Feature importance ranking plays a role too. Saliency analysis of geometric constraints fits in there as well. The interpretability score works out to the percentage of decision steps that an engineer can trace using those tools.

Moreover, qualitative metrics will be considered, such as the interpretability level of the model (measured as the ability of the engineer to comprehend the decision reasoning of the AI system) as well as the verification of spatial and topological relationships).

These considerations will be brought into compliance with established international norms, such ISO 1101 for geometrical product specification, as well as with ISO 286 for dimensional tolerances. Moreover, parameters relevant for robust assessment under conditions of strong variability will be incorporated, such as stability ratings or measures of algorithms' confidence (Fig. 1). Such issues provide a robust method of engineering verification that would apply to the real-world application of AI-based design.

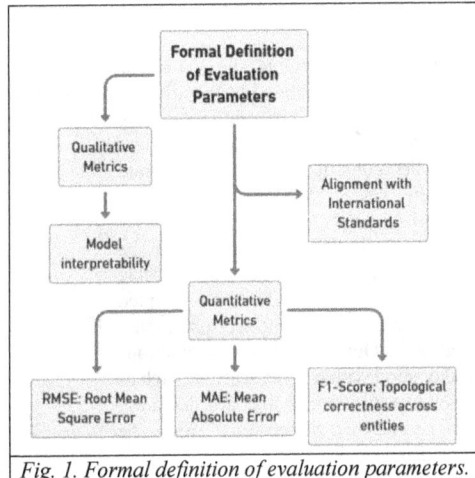

Fig. 1. Formal definition of evaluation parameters.

Development of testing scenarios

This stage will focus on evaluating the performance of the AI algorithms in different, controlled, and repeatable scenarios. In this proposal, there will be the development of reference models that will cover the widest possible range of engineering tasks, such as standard geometries as well as more complex, unusual models that would be difficult or impossible to generate through conventional CAD systems. Each model will come with a well-defined set of geometric, functional, and dimensional specifications.

Right now, the proposed standardized CAD plus AI test set lacks any unified benchmark. We suggest creating it anyway. This new set needs to cover native CAD representations in B-Rep format. It also requires full parametric history trees. Feature-level annotations count too, things like sketches, constraints, and Boolean operations. Finally, pair up ground-truth models with ones generated by AI.

Scenarios will also comprise edge cases as those with a discontinuous boundary, partial symmetry, or a degenerated feature, which often give rise to failures in automated processing.

Large parametric variability will be considered, with models being varied through variations in input variables (length, angle, diameter) in order to test robustness (Fig. 2).

With a focus on the issues of reproducibility as well as enabling comparisons between different institutions, the proposal includes the introduction of a standardized CAD+AI test set.

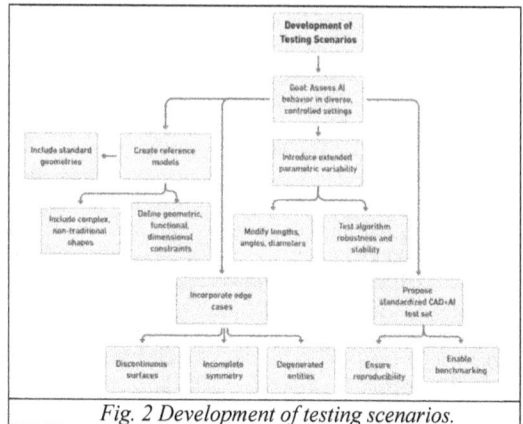

Fig. 2 Development of testing scenarios.

Application of computational and statistical analysis tools

At this stage, the defined tests will be run on a sufficiently large set of automatically generated instances, which will measure the predictability as well as robustness of the predictions (Fig. 3).

Each test will be carried out under controlled conditions with uniform input, and the results will be saved in a centralized database.

Reproducibility of results will be achieved through the use of the same dataset, algorithms with fixed parameters, and uniform processing platforms. During analysis, calculations of deviation with robust measures such as ensemble spread and standard deviation of geometric dimensions will be considered.

Error distribution will be statistically analyzed through multivariate regression analysis, cluster analysis, as well as other visualization tools such as scatter plots and density plots, in order to see the correlation between performance and input properties, group models with

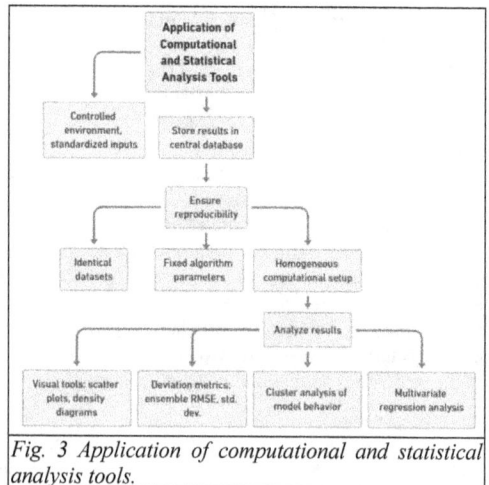

Fig. 3 Application of computational and statistical analysis tools.

similar patterns, as well as examine regions where the AI model performs well and regions where improvement of the algorithms is needed.

Error mapping and determination of algorithmic limits

At this stage, the task is focused on creating a comprehensive error profile for every type of geometry that the algorithm will work with. Error profiles will contain numerical information

Innovative Manufacturing Engineering and Energy - IMANEE2025   Materials Research Forum LLC
Materials Research Proceedings 61 (2026) 77-83   https://doi.org/10.21741/9781644903995-10

about defects, such as dimensional, topological, or connectivity errors that might occur, along with their frequencies (Fig. 4).

Whether the errors are systematic, recurrent, or random will be investigated, and correlations will be established between the aforementioned deviations and input variables such as complexity, geometric entities, or type of constraints.

To facilitate the interpretation of algorithms across the world, the proposal is the introduction of a composite indicator of algorithmsic risk. Such a measure will be determined through a weighted process that considers different variables, such as the intensity of the error, their density, as well as the value in project functionality.

The indicator will facilitate both the classification of models based on their associated confidence level as well as comparisons between different versions/algorithms. Thus, the mapping stage will play a critical role in

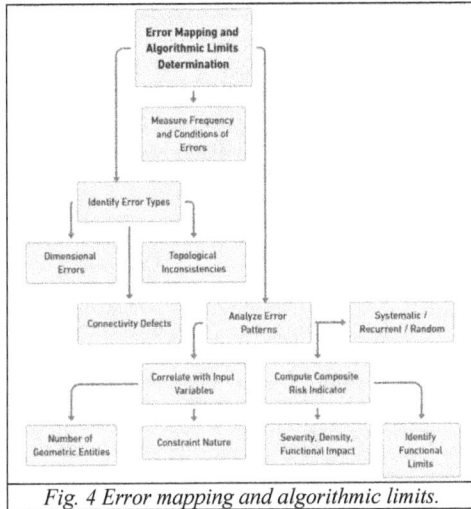

Fig. 4 Error mapping and algorithmic limits.

determining the functional boundaries as well as optimizing based on where such boundaries are most needed.

In additive manufacturing, a deviation in a geometric feature corresponds to a particular failure mode. For example:

- Thickness errors: risk of under-extrusion (FDM) or incomplete fusion (SLM).
- Overhang deviation: support failure and collapse.
- Sharp topological discontinuities: thermal warping or residual stress concentration.

Thus, the error maps are directly translated into AM failure likelihoods, strengthening the relevance of the proposed framework to manufacturing applications.

Formulation of applicable recommendations

This last stage of methodology merges the insights acquired from previous evaluations, thus giving proper guidance on the optimization of the AI system. Following the conclusions drawn from the work of Chen & Li [5], that indicated the importance of adapting the loss function as well as incorporating human feedback into the generation task, three overarching paths will be considered.

A composite risk indicator based on criticality levels identified in ISO 1101 and ISO 286 standards can be expressed as:

$$R_{total} = w_1 \cdot E_{dim} + w_2 \cdot E_{topo} + w_3 \cdot I_{func} \tag{1}$$

where:

E_{dim} – aggregated dimensional deviation (e.g., RMSE),

E_{topo} – topological inconsistency index,

I_{func} – functional sensitivity score (impact on manufacturability),

w_1, w_2, w_3 – weights derived from tolerance classes (fine/medium/coarse).

Nonlinear weightings (exponential penalties) may be considered for critical AM features such as thin walls and overhanging structures.

Primarily, modular adaptability of neural net models is recommended, thereby enabling every system component to focus on a well-delineated set of engineering tasks, such as processing functional geometry information or imperatives related to dimensional tolerances. Modularization is further aided by recent studies on explainable AI in industrial applications [6].

Second, the use of active feedback from the design engineer in the learning process is recommended. This may be achieved through reinforcement learning or human guidance. This is recommended because recent studies indicate that human involvement in the loop of algorithms enhances solution quality [7]. Thus, not only will this approach optimize predictions, but controlling AI decisions will also be ensured (as seen in Fig. 5).

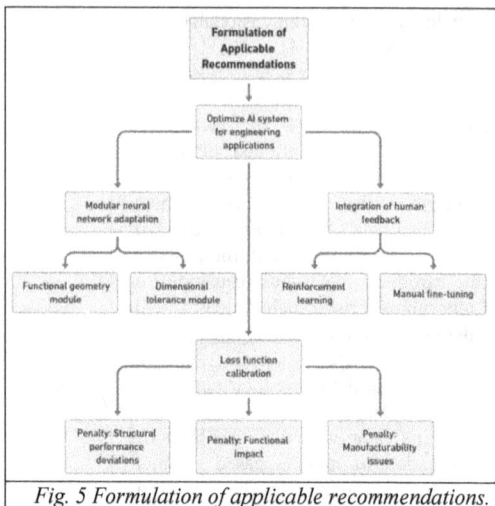

Fig. 5 Formulation of applicable recommendations.

Third, the importance of calibrating the loss functions, based on their application specifics, is highlighted. Loss functions must correctly treat the magnitude of imperfections that may impact structure, functionality, or manufacturability. Such studies, as carried out by Bui [8], show the importance of adapted loss functions in boosting the geometric precision of design models produced by AI. Therefore, this stage becomes a crucial element in tailoring the AI model for engineering tasks with stringent performance requirements [10-11].

They incorporate human feedback through Reinforcement Learning from Human Feedback, or RLHF. In that setup, experts make corrections to sketches, constraints, or features. Those corrections serve as the reward signals. On top of this, interactive constraint editing lets engineers guide the generation paths right in real time.

Conclusions

The paper focuses on the significance of test scenarios that are reproducible and formulated based on CAD models differing both geometrically and topology-wise. The significance of such test scenarios emerges from the fact that only through exposure of algorithms to such data can one witness the reaction of the algorithm towards parameters in areas known from the literature as responsible for common errors in automatic generation.

Among the strengths of this research work is the standardization of parameters in both quantitative and qualitative evaluations such as those in RMSE, MAE values, topologically adapted F1 scores, interpretability scores, and parameters for correctness in structure. These parameters are based upon relevant ISO guidelines for geometrical product specifications and are thus effective in terms of technical verification of AI-developed models. The methodology developed here incorporates parameters more aligned with accuracy/robustness in terms of model prediction fidelity and reliability.

The standard CAD+AI system is one of the most important components of the proposed framework because this system provides the capability of systematically assessing the performance of various algorithms under different geometrical, topological, and parametric constraints in such a way that the experimental results are compared in a similar context. The combination of different advanced statistical analysis techniques like multivariate regression analysis, overall deviation analysis, behavior clustering analysis, and mapping of deviation distribution provides in-depth insights regarding the working of the various algorithms.

The proposed framework can be adapted to various scenarios in the industry and is scalable and reproducible. The proposed research not only provides key contributions in terms of applications in the industry but also marks the beginning of a milestone in theory in providing clarity regarding the required parameters in assessing AI in the context of CAD and paving the way towards the formation of future international standards in this realm.

References

[1] I. Goodfellow, Y. Bengio & A. Courville, *Deep Learning*. MIT Press, 2016.

[2] A. Géron, *Hands-On Machine Learning with Scikit-Learn, Keras, and TensorFlow: Concepts, Tools, and Techniques to Build Intelligent Systems*. O'Reilly Media, 2019.

[3] S. Russell & P. Norvig, *Artificial Intelligence: A Modern Approach* (4th Edition). Pearson, 2020.

[4] Y. Zhao, X. Gao & Z. Lin, *A Review of Deep Learning Techniques for Additive Manufacturing*. Journal of Manufacturing Processes, 77, 375–389. doi.org/10.1016/j.jmapro.2022.03.017, 2022.

[5] Y. Chen & Y. Li, *Design automation for additive manufacturing using generative design and machine learning*. Computers in Industry, 2021.

[6] Y. Zhang, L. Sun, C. Wang & Z. Liu, *Modular Neural Network Architectures for Engineering Design Automation*. Advanced Engineering Informatics, 2022.

[7] A. P. Schoellig, A. Tsiamis, M. Müller & A. Papachristodoulou, *Human-in-the-loop AI for engineering design: A survey of recent advances*. Annual Reviews in Control, doi.org/10.1016/j.arcontrol.2020.06.006, 2020.

[8] H. Bui, D. Phung, D. Do & B. Vo, *Loss Function Engineering for High-Precision 3D Geometry Generation in Industrial Design*. Computer-Aided Design, 136, 103030. doi.org/10.1016/j.cad.2021.103030, 2021.

[9] G. Cheng, S. Liu & Y. Wu, *AI-Driven Topology Optimization: Advances and Challenges*. Structural and Multidisciplinary Optimization, 67(4), 2023.

[10] Y. Wang, Y. Sun, Z. Liu, S. Sarma, M. Bronstein & J. Solomon, *Dynamic Graph CNN for Learning on Point Clouds*. ACM Transactions on Graphics, 35, 2019. https://doi.org/10.1145/3326362

[11] L. Mescheder, M. Oechsle, M. Niemeyer, S. Nowozin & A. Geiger, *Occupancy Networks: Learning 3D Reconstruction in Function Space*. CVPR 2019. https://doi.org/10.1109/CVPR.2019.00459

Innovative Manufacturing Engineering and Energy - IMANEE2025   Materials Research Forum LLC
Materials Research Proceedings 61 (2026) 84-91   https://doi.org/10.21741/9781644903995-11

3D printed rPETG applications for interior design: A case study

Lazaros FIRTIKIADIS[1,a *], Panagiotis KYRATSIS[1,b], Nikolaos EFKOLIDIS[1,c]

[1]University of Western Macedonia, Department of Product and Systems Design Engineering, Kila Kozani, GR50100, Greece

[a]l.firtikiadis@uowm.gr, [b]pkyratis@uowm.gr, [c]nefkolidis@uowm.gr

Keywords: Parametric Design, Customized Products, 3D Printing, Recycled PETG, Circular Economy

Abstract. The use of recyclable filaments to produce 3D models using the Fused Filament Fabrication (FFF) method is increasingly promoting sustainability and innovation, compared to traditional filaments. The continuous improvement of the characteristics of this type of filament contributes to the control of their mechanical properties and especially their strength. At the same time, their compatibility with existing 3D printing technologies encourages their integration into the manufacturing sector. Polyethylene terephthalate glycol (rPETG) is one of the most popular recyclable materials in recent years. In addition to promoting the circular economy, it also offers new possibilities in product manufacturing. It can be used as a material for the 3D fabrication of products such as household items and office supplies. The present work focuses on the design and fabrication of handles used in cabinets and drawers through the parametric design of a kitchen island, highlighting the choice of rPETG as suitable material for their fabrication. The aim is for the final everyday object to be adaptable to different types of cabinets. In its own way, it offers ease, comfort, and safety while emphasizing the overall style of the furniture and harmonizing with the rest of the space.

Introduction

Parametric design is a rapidly evolving design methodology that is increasingly being applied in industry [1]. The advantages for designers include the ability to create complex and alternative product shapes through the use of algorithms and advanced CAD (Computer-Aided Design) systems [2,3]. The parameter values selected by the designer, or sometimes even by the user, dictate the resulting geometry each time. In fact, this particular design method offers the opportunity for designers to improve the final result of their idea compared to more traditional design methods [4-6].

3D printing is one of the modern manufacturing technologies used in the manufacturing of products. Fused filament fabrication (FFF) is a 3D printing technology that provides the ability to manufacture customized household products according to the needs of the user himself. In this way, adaptability is enhanced, leading to a more targeted and personal product design for consumers [7].

At the same time, integrating 3D printing with 3D design software allows for the verification of components prior to manufacturing, leading to enhanced outcomes. The 3D printing methodology is increasingly included in the specific design processes, enhancing the final result of the product [8]. Additionally, each time it is deemed necessary to choose the most appropriate construction material in order to produce either the prototype model for testing or the final product [9-12].

Innovative Manufacturing Engineering and Energy - IMANEE2025   Materials Research Forum LLC
Materials Research Proceedings 61 (2026) 84-91   https://doi.org/10.21741/9781644903995-11

A Computer-aided Design approach to a Kitchen Island application

It is crucial to highlight that parametric design is increasingly being incorporated into the field of interior design. This approach offers new possibilities and perspectives for furniture design within residential interiors, enhancing the overall design process.

The kitchen is one of the most frequently used areas in a home. An efficient layout enhances both ergonomics and functionality, providing greater comfort and safety for its occupants. One method to achieve this is to develop mathematical correlations. By considering the total surface area of the space in relation to the proposed layout, the designer can utilize data to enhance the effectiveness of their design solution [13]. In addition, parametric design allows for the development of alternative kitchen furniture proposals, such as the island. The definition of parameters can modify the shape of the final furniture according to the users' choices [14].

Simultaneously, the use of parametric tools enables the optimization of the final model, as predefined constraints are defined from the beginning of the design process. In particular, the integration of parametric tools with BIM (Building Information Modeling) platforms further enhances the assembly process of the final model, significantly reducing both production time and cost [15-17]. Finally, the selection of appropriate materials can contribute to sustainability and innovation in the design and construction of kitchen furniture [18].

In this study, the proposed methodology of parametric design was followed, using the Rhinoceros™ 7 software and the Grasshopper™ application. The main purpose of the aforementioned case study is to highlight the parametric design usage in the whole design procedure of a kitchen island application for interior spaces. The figure below (Fig. 1) represents the design process, organized in the form of a flowchart, with the use of appropriate symbols.

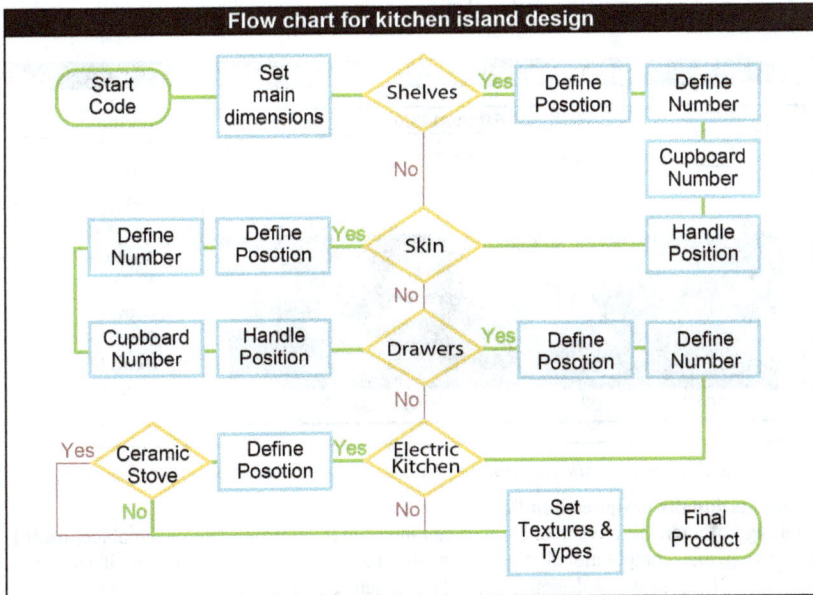

Fig. 1: The basic stages of kitchen island design, illustrated in a flowchart.

Innovative Manufacturing Engineering and Energy - IMANEE2025   Materials Research Forum LLC
Materials Research Proceedings 61 (2026) 84-91   https://doi.org/10.21741/9781644903995-11

Specifically, the design process starts by waiting for the end-user to input the fundamental dimensions of the kitchen island furniture. Depending on the selected length, the corresponding number of cabinets that will make up the final piece of furniture is automatically determined. It is worth noting that if the final length of the furniture is greater than the sum of the boxes, then an additional blank cabinet is automatically added to cover the resulting gap, for both functional and aesthetic reasons. Next, the end user will be able to select which cabinets will be included in the final result and specify their placement order. The cabinets have been fully developed using parametric design and are categorized within a specific furniture ontology, which includes a) shelves, b) sink, c) drawers and d) electric stove with hob. Moreover, the number of shelves can be defined, while they can also include cabinet doors in front. The end-user can select, based on their preferences and aesthetic considerations, whether to have one or two doors, and choose the side on which the handle is positioned, ensuring the door opens in the desired direction.

In the same way, the procedure applies to the cabinet in which the sink with the faucet is integrated. Accordingly, the number of drawers is defined separately. Regarding the electric stove, the user can choose whether a ceramic hob will be installed on the island counter.

Finally, in order to offer even more choices to the customer, a list of the individual items of the island was created. This list includes a series of different handles for the cabinets and drawers, different styles of faucet, ceramic hob, as well as different materials from which the individual boxes and the counter of the island will be made.

Fig. 2 illustrates an example representation of a study, which results in the kitchen island formed by the customer's choices, as presented in the aforementioned flow chart of Fig. 01 the purpose of the final product is to fulfill the customer's requirements for their kitchen space while aligning with their aesthetic preferences. The photorealistic rendering visualizes the example of this study (Fig. 2).

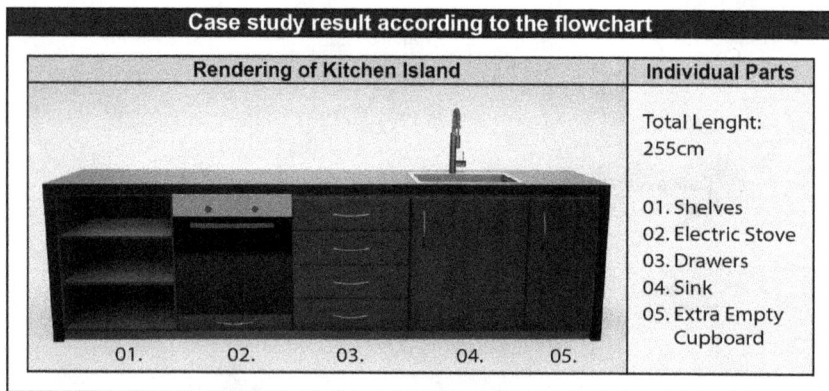

Fig. 2: Photorealistic illustration of a case study based on the above flowchart.

A computationally designed handle

The design of alternative types of handles can improve the functionality of the object itself [19]. Considering that computational design can also be used for the automation of smaller-scale products [20], it allows to design a series of kitchen furniture handles in a more efficient and precise manner.

The proposed code developed in the Grasshopper™ software. The main goal of the suggested design application is the production of parametric handles for drawers and cabinet doors. More specifically, the designer/end-user, by setting different values in the geometry parameters, has the

Innovative Manufacturing Engineering and Energy - IMANEE2025   Materials Research Forum LLC
Materials Research Proceedings 61 (2026) 84-91   https://doi.org/10.21741/9781644903995-11

ability to create a family of round handles with different morphological characteristics. The diversity of the results allows the creation of a range of final products that can be produced, combining aesthetics with functionality.

The key parameters that can be modified through the design code enable the instant modification of the handle's geometry. Firstly, the height and thickness of the handle can be adjusted. By defining specific control points, the desired distances are achieved, as well as the diameter of the corresponding circular areas. To enhance the aesthetic appeal of the final product, a structure is formed, driven by the irregular shape of the corals. Adjusting the parameter values also results in a different casing design. In addition, to further enhance the visual interest of the resulting geometry, the rotation parameter was also incorporated. Finally, another basic element that enhances the complexity of the design is the shape of the end that the specific product can have (from flat to spherical geometry).

Key parameters controlled include the height, thickness, diameter of circular sections, and the rotation of the handle. The process relies on fundamental CAD commands, such as Circle and Sphere to create basic geometric shapes, Explode and Rebuild Curve to control and modify the handle's curve form, Plane Normal and Rotate for orientation and rotation of the geometry, Cap Holes and Solid Intersection for managing mounting holes and merging various geometric elements, Contour and Dispatch to define boundaries and distribute elements of the handle, Smooth Curve and Momentum for aesthetic enhancement and smooth flow of shapes, as well as Arithmetic Mean and Spring used for dynamic adjustment of control points that determine the shape. By using these commands, the parametric model enables quick and precise adaptation of the handles to fit different dimensions and functional needs while ensuring their proper integration into furniture, as demonstrated in the examples of Fig. 3. This methodology provides a clear structure and flexibility, guaranteeing design accuracy and immediate adaptability throughout the handle development process.

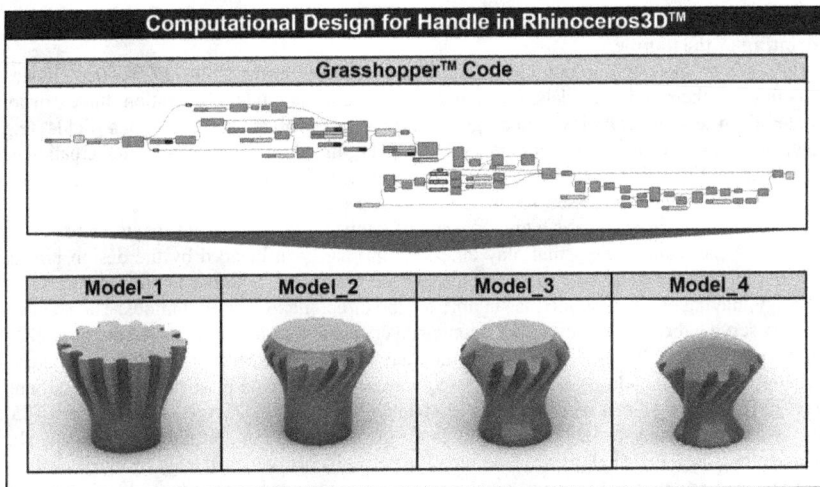

Fig. 3: Customized handles based on design code.

In addition to illustrating the handle itself, it is necessary to show it installed on a cabinet door or drawer. For this reason, a series of alternative handles was chosen to be placed in the kitchen island case study, which was presented earlier. The photorealistic depiction (Fig. 4) of the handle's highlights both the individual aesthetics of the product and its appearance when applied to kitchen furniture. In addition, the designer has the ability to incorporate the prospective customer's preferences into the final form of the handle and to make any necessary changes before production.

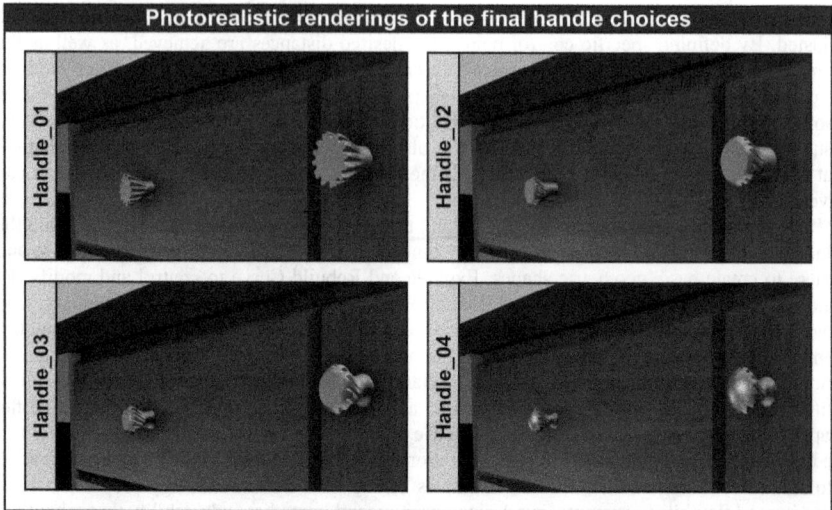

Fig. 4: Photorealistic illustration with the placement of handles on the cabinets of the kitchen island case study.

3D printing of the handle

3D printing is an innovative production technology which offers the ability to produce physical objects through their digital models. By taking advantage of its mode of operation, the extrusion of material in layers, complex object geometries can be obtained, in various fields (e.g. architecture, industry, arts, sciences, wearables) [21]. In addition, it allows the creation of personalized products that meet the needs of the users [22]. This advantage differentiates it from conventional production processes, reducing both cost and production time.

In industry and market there is a great variety of knobs used for different functions (e.g. cars, aircraft). The particular variety that may emerge is primarily influenced by the design process within digital three-dimensional modeling programs. Moreover, additive manufacturing (AM) technology, having evolved rapidly, is suitable for the direct manufacture of knobs. The resulting object can serve either as a prototype for geometry verification, or as the final product itself [23]. An additional element that can help optimize the final product is the appropriate choice of material for its construction. In addition to the most used materials, reinforced plastics are now also being chosen [24]. The most common material used in 3D printing is PLA (polylactic acid) [25]. However, there are other widely used materials for the production of 3D functional objects. One of them is recycled PETG (Polyethylene terephthalate glycol), which thanks to its mechanical properties gives an innovative and sustainable character to the products produced [26,27].

In this study, the 3D printed geometries were manufactured on the CreatBot D600 Pro 3D printer. The specific 3D printer belongs to the FFF (Fused Filament Fabrication) category, with allowed printing dimensions of 600x600x600mm. The material chosen for the prints is NEEMA3D™ rPETG. For the optimal printing result of the parametric handles, specific printing settings were applied. More specifically, and according to the study [28] the parameters that were set were the filling percentage (90%), the printing speed (20mm/s) and the type of filling (Wiggle). These parameters were configured and fine-tuned using the Simplify3D^R software to ensure precise control of the printing process. The selection of these specific parameters increases the tension, which influences the object's mechanical strength, since the fundamental motion of opening and closing a cabinet or drawer places stress on the handle. The table below, Table 1, lists all the parameters that were set during the 3D printing.

Table 1: 3D printing parameters

Operation Parameter	Value
Infill	90%
Speed	20 [mm/s]
Infill Type	Wiggle
Nozzle Temperature	240 [°C]
Bed Temperature	60 [°C]
Filament Diameter	1.75 [mm]
Nozzle Diameter	0.6 [mm]
Layer	0.2 [mm]
Flow	100%
Cooling Fan	100%

Finally, Fig. 5 presents the stages of 3D printing of the handles selected during the parametric design process.

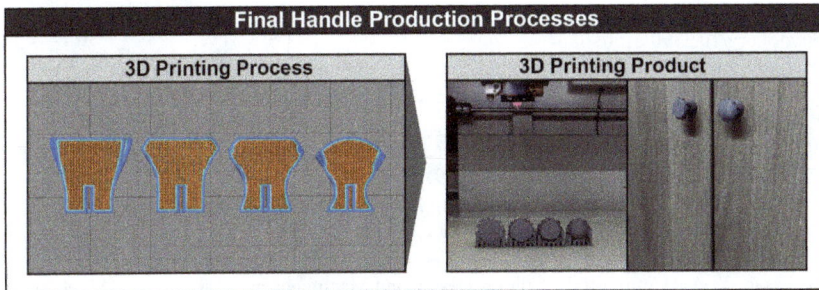

Fig. 05*: The steps for 3D printing the handles.*

Conclusions

Considering both the island and handle design methodology, the countless possibilities that parametric design can offer are highlighted. It manages to overcome obstacles that designers would potentially encounter in the design phase. The algorithmic approach can offer organization, simplification and efficiency in design, greatly improving the holistic process. Designers are given the opportunity to create product families, thanks to the diversity that can arise. The definition of specific parameters by the designer himself or even the customer himself, results in the final product meeting the aesthetic preferences as well as the functional needs of each customer. In this way, personalization contributes to user satisfaction, while ensuring the uniqueness of the product, gaining even greater importance for the customer.

At the same time, 3D printing technology enables the creation of unusual, complex, and personalized geometries generated through parametric design. Thanks to the advantages offered by 3D printing, it becomes an indispensable tool for designers.

This specific research, with the development of two case studies, aims to highlight parametric design in the field of interior decoration, and more specifically in the area of the domestic kitchen.

The design methods that follow in both cases present the diversity that can arise in a kitchen island furniture, but also in a type of handle for its cabinet doors and drawers. This new trend in design and industry has resulted in the creation and production of customized products, which are created by both the designer and the end-user. The participation of each customer in determining the characteristics of the product of their choice enhances the uniqueness and adaptability of their final products. Finally, the development of interdisciplinarity, which includes the collaboration of design and technology, creates more and more prospects for the development of product personalization in the field of interior decoration.

References

[1] N. El-Mabrouk, Predicting the Evolution of Syntenies-An Algorithmic Review, Algorithms 14(5), (2021) 152. https://doi.org/10.3390/a14050152

[2] I. Volkau, S. Krasovskii, A. Mujeeb, H. Balinsky, A Non-Gradient and Non-Iterative Method for Mapping 3D Mesh Objects Based on a Summation of Dependent Random Values, Algorithms 17(6) (2024) 248. https://doi.org/10.3390/a17060248

[3] P. C. Gembarski, Joining Constraint Satisfaction Problems and Configurable CAD Product Models: A Step-by-Step Implementation Guide, Algorithms 15(9) (2022) 318. https://doi.org/10.3390/a15090318

[4] A. Köhler, M. Breuß, Computational Analysis of PDE-Based Shape Analysis Models by Exploring the Damped Wave Equation, Algorithms 15(9) (2022) 304. https://doi.org/10.3390/a15090304

[5] Y. Li, N. Zhang, Y. Gong, W. Mao, S. Zhang, Three-Dimensional Elastodynamic Analysis Employing Partially Discontinuous Boundary Elements, Algorithms 14(5) (2021) 129. https://doi.org/10.3390/a14050129

[6] P. Kyratsis, E. Gabis, A. Tzotzis, D. Tzetzis, K. Kakoulis, CAD based product design: A case study, Int. J. Mod. Manuf. Technol 11(3) (2019) 88-93.

[7] P. K. Jain and P. K. Jain, Use of 3D printing for home applications: a new generation concept, Materials Today: Proceedings 43 (2021) 605-607. https://doi.org/10.1016/j.matpr.2020.12.145

[8] K. Erşan, Y. A. Demiroğlu, B. Güldür, Design and Manufacturing of Real-Scale-Mockup-Car Door Via 3d Printer, Journal of New Results in Science 7(3) (2018) 54-66.

[9] J. Kechagias, K. Kitsakis, A. Zacharias, K. Theocharis, K. E. Aslani, M. Petousis, N. A. Fountas, N. M. Vaxevanidis, Direct 3D Printing of a hand splint using Reverse Engineering, In IOP Conference Series: Materials Science and Engineering 1037(1) (2021) 012019. https://doi.org/10.1088/1757-899X/1037/1/012019

[10] C. I. Rizescu, D. E. Baciu, D. Rizescu, Study behaviour of ratchet mechanism manufactured with 3D printing technology, In IOP Conference Series: Materials Science and Engineering 1235(1) (2022) 012001. https://doi.org/10.1088/1757-899X/1235/1/012001

[11] D. Rizescu, D. Besnea, C. Rizescu, E. Moraru, I. Panait, Rapid prototyping method for replacing non-circular gears, In IOP Conference Series: Materials Science and Engineering 1235(1) (2022) 012005. https://doi.org/10.1088/1757-899X/1235/1/012005

[12] I. Anderson, Mechanical properties of specimes 3D printed with virgin and recycled polylactic acid, 3D Print. Addit. Manuf. 4 (2017) 110-115. https://doi.org/10.1089/3dp.2016.0054

[13] D. A. Yazıcıoğlu, A statistical data analysis for increasing the kitchen design performance, A| Z ITU Journal of the Faculty of Architecture 11(1) (2014) 174-184.

[14] X. Sun, X. Ji, Parametric model for kitchen product based on cubic T-Bézier curves with symmetry, Symmetry 12(4) (2020) 505. https://doi.org/10.3390/sym12040505

[15] C. Dautremont, S. Jancart, C. Dagnelie, A. Stals, Parametric design and BIM, systemic tools for circular architecture, In IOP Conference Series: Earth and Environmental Science 225(1) (2019) 012071. https://doi.org/10.1088/1755-1315/225/1/012071

[16] J. Park, BIM-based parametric design methodology for modernized Korean traditional buildings, Journal of Asian Architecture and Building Engineering 10(2) (2011) 327-334. https://doi.org/10.3130/jaabe.10.327

[17] W. Wahbeh, Building skins, parametric design tools and BIM platforms. In Conference Proceedings of the 12th Conference of Advanced Building Skins (2017) 1104-1111.

[18] M. Doiro, F. J. Fernández, M. J. Félix, G. Santos, Machining operations for components in kitchen furniture: A comparison between two management systems, Procedia Manufacturing, 41 (2019) 10-17. https://doi.org/10.1016/j.promfg.2019.07.023

[19] A. Abouhossein, Z. Ansari, S.R. Kalkhoran, 3D Printed Hands-Free Door Handle to Prevent COVID-19 Virus Spread: Developing a Design Based on Iranian Population for Medical Centers, Human Aspects of Advanced Manufacturing, Production Management and Process Control 146(146) (2024). https://doi.org/10.54941/ahfe1005163

[20] A.G. González, J. García Sanz-Calcedo, O. López, D.R. Salgado, I. Cambero, J.M. Herrera, Guide Design of Precision tool Handle Based on Ergonomics Criteria Using Parametric CAD Software, Procedia Engineering, 132 (2015) 1014-1020. https://doi.org/10.1016/j.proeng.2015.12.590

[21] P. LIGKA, N. EFKOLIDIS, A. MANAVIS, P. KYRATSIS, Manufacturing computationally designed wearables via 3D printing.

[22] A. Manavis, P. Minaoglou, K. Aidinli, N. Efkolidis, P. Kyratsis, CAD based design for the jewellery industry: A case study, In IOP Conference Series: Materials Science and Engineering 1009(1) (2021) 012036. https://doi.org/10.1088/1757-899X/1009/1/012036

[23] M. S. A. Samsaimon, M. S. Wahab, The Design of Door Handles for Civil Aircraft Cabin Doors Utilizing Addictive Manufacturing for Aerospace Component Manufacturing, Progress in Aerospace and Aviation Technology 3(1) (2023) 9-13. https://doi.org/10.30880/paat.2023.03.01.002

[24] G. N. Raut, V.P. Mali, Optimizations of outer door handle in automobiles using composite material (2017)

[25] A. Lang, Experimental characterisation of physical properties of different PLA filaments used in FFF 3D printing technology, In IOP Conference Series: Materials Science and Engineering 1235(1) (2022) 012002. https://doi.org/10.1088/1757-899X/1235/1/012002

[26] A. K. Singh, R. Bedi, B.S. Kaith, Composite materials based on recycled polyethylene terephthalate and their properties-A comprehensive review, Compos. Part B Eng., 219 (2021) 108928. https://doi.org/10.1016/j.compositesb.2021.108928

[27] M. Kováčová, J. Kozakovičová, M. Procházka, I. Janigová, M. Vysopal, I. Černičková, J. Krajčovič, Z. Špitalský, Novel hybrid PETG composites for 3D printing, Applied Sciences, 10(9) (2020) 3062. https://doi.org/10.3390/app10093062

[28] L. Firtikiadis, A. Tzotzis, P. Kyratsis, N. Efkolidis, Response Surface Methodology (RSM)-Based Evaluation of the 3D-Printed Recycled-PETG Tensile Strength, Applied Mechanics, 5(4) (2024) 924-937. https://doi.org/10.3390/applmech5040051

Keyword Index

About the Editor

Associate Professor Marius-Andrei Mihalache is an established academic and researcher in advanced manufacturing engineering, affiliated with the Gheorghe Asachi Technical University of Iasi (TUIASI). His career is distinguished by a sustained commitment to teaching, scientific research, and technological innovation in manufacturing systems and processes. He holds roles in academic instruction, research supervision, and international scholarly exchange within mechanical and industrial engineering domains. His pedagogical approach integrates theoretical foundations with practical applications, equipping students with both conceptual understanding and hands-on skills essential for modern manufacturing environments.

Dr. Mihalache's research portfolio centers on integrated manufacturing technologies and digital transformation in production systems. Key research thrusts include: optimization of design and industrial product development through selective deployment of CAD/CAM/CAE/CAx/PDM/PLM software systems, technological process optimization leveraging software tools, and detailed finite element analysis and development of automation solutions for manufacturing process workflows. His scholarly work contributes to advancing knowledge and improving performance in manufacturing design, process planning, and digital manufacturing implementation. Dr. Mihalache has contributed to multiple peer-reviewed conference proceedings and scientific publications with an impact factor, both as author and reviewer.

His research is indexed in major academic databases, including Web of Science and Scopus, reflecting international recognition and citation of his work in core engineering disciplines. Dr. Mihalache actively participates in academic leadership and collaborative activities, such as advisement and supervision of the undergraduate and graduate student projects, engagement in international academic mobility programs (e.g., Erasmus+ partnerships with universities in Poland), as well as ongoing involvement with the IManEE conference series, thus contributing to scholarly review processes and scientific community build-up.

Dr. Marius Andrei Mihalache's sustained involvement in research, teaching, and international academic forums positions him as a respected contributor to the global manufacturing engineering community. This would not be possible without the sense of community, which favors innovation within research and teaching activities, complemented by a united and performance-driven department. His commitment in the field of Industrial Engineering sustains focus on advancing both personal expertise and institutional performance through continuous engagement in teaching, research, and academic service. His professional trajectory reflects a strong dedication to lifelong learning, methodological rigor, and the integration of emerging scientific and technological developments into engineering education and research practice.

Dr. Marius Andrei Mihalache actively seeks new opportunities for national and international collaboration, recognizing the strategic value of interdisciplinary and cross-institutional partnerships in addressing complex industrial and societal challenges. His interests include joint research projects, co-authored publications, participation in academic networks, and the development of innovative curricula aligned with contemporary industrial needs.

Within the department of Machine Manufacturing Technology from the Faculty of Machine Manufacturing and Industrial Management, he maintains an active and constructive role, contributing to academic governance, program development, and improved research capacity. He is particularly motivated to support a collaborative and inclusive academic environment, where knowledge sharing and mutual professional growth are encouraged. On a human and mentorship level, he is deeply invested in guiding early-career academics and young researchers, assisting them in developing research competencies, teaching skills, and academic visibility. Through supervision, informal mentoring, and collaborative work, Dr. Marius Andrei Mihalache supports young colleagues in their pursuit of higher levels of expertise, professional recognition, and career advancement.